I0798747

A HISTORY OF THE ANDOVER IRONWORKS

Come Penny, Go Pound

KEVIN W. WRIGHT

THE History PRESS

Published by The History Press
Charleston, SC 29403
www.historypress.net

Copyright © 2013 by Kevin W. Wright
All rights reserved

Front cover: Old Andover Gristmill, postcard view, circa 1910.
Back cover: *Courtesy Chester County Historical Society.*

First published 2013

Manufactured in the United States

ISBN 978.1.62619.218.8

Library of Congress CIP data applied for.

Notice: The information in this book is true and complete to the best of our knowledge. It is offered without guarantee on the part of the author or The History Press. The author and The History Press disclaim all liability in connection with the use of this book.

All rights reserved. No part of this book may be reproduced or transmitted in any form whatsoever without prior written permission from the publisher except in the case of brief quotations embodied in critical articles and reviews.

Contents

Acknowledgements

I gratefully acknowledge my sources of inspiration and information. Before all others, I thank my wife, Deborah Powell, who contributed considerable time and talent to the preparation of the illustrations and manuscript. And no one has been more persistent, or more encouraging, over the decades than my friends from the Canal Society of New Jersey, Bill Moss and Robert Goller. I am also indebted to the archivists at the Historical Society of Pennsylvania, who provided access to the Chew Family Papers, a most valuable resource for anyone interested in the Andover Ironworks. These papers were not yet catalogued when I studied them in 1981. Special thanks to Andrea Ashby, library technician at Independence National Historical Park, for her assistance in obtaining a digital copy of the portrait of William Allen by Robert Feke (1746) and to Ellen E. Endslow, director of collections/curator for the Chester County Historical Society, West Chester, Pennsylvania, for her assistance in obtaining a digital copy of the portrait of Joseph Turner (circa 1752).

For most of my life, I have enjoyed Bill Gibbs's historical drawings of the old Andover Barn and Gristmill, which first hung on the walls of my grandparents' home and later in my own house—I thank him for allowing me to reproduce his 1968 rendering of the old Andover Barn.

For shaping my interest and honing my craft, I acknowledge my family, mentors and colleagues, especially V. Ivan Wright, Gertrude Brink Wright, Mary G. Mullen, Teresa M. Wright, Alex Everitt, Jim and Mary Lee, Claire Tholl, Betty Schmelz, Dr. Richard Lenk, John Spring, Bert Prol, Paul

Taylor, Beverly Weaver, Lou Cherepy, Todd Braisted and Reg McMahon. For caring and sharing the story of their interesting family, I recognize the contributions of Sandford Roy Smith and Robert H. Smith. For special research assistance, I thank Richard M. Stevens and Barbara Waskowich. Ivan, Ben and Anna have been constant, interesting and encouraging companions on my historical journey.

Lastly, to all the women and men who lived this history (who I sometimes feel as if I know), though we will never meet, you have my admiration and respect. May these pages bring you a taste of immortality!

Introduction

William Allen, Joseph Turner, Lynford Lardner and John Hackett purchased New Jersey's largest hematite-magnetite deposit in 1759, naming their iron company for Turner's birthplace in Andover, England. Starting in 1761, ore was smelted at a charcoal blast furnace in Andover Borough, Sussex County, New Jersey, and refined into wrought-iron bars, seven miles distant, at Andover Forge on the Musconetcong River. Old Andover Forge was renamed Waterloo in 1840. Although the high-grade ore was especially suited to the production of blister steel—used to make edge tools—the Andover Ironworks closed in 1795. With only horseflesh and oxen to gather raw materials and convey output to tidewater across rough miles of rocky wilderness, many Highland charcoal-iron plantations, such as Andover, flickered out once the primeval forest in their proximity was cut over for fuel.

This is the first full telling of the history of this particular eighteenth-century American ironworks, regarded as one of the most important in its day. To better illuminate the subject matter for a lay audience (including myself), I describe the various crafts and processes of manufacture associated with such water-powered industrial plantations, hopefully with only enough detail to sustain interest. Since this technology was somewhat standard in its day and generally reflected the latest European innovations, this work may hold a more general interest than its locality suggests.

Genealogy and a strong sense of place drew me to this subject matter more than thirty years ago. My grandfather Ivan Wright's family descends from the iron men and women of the Highlands, where the names of Wright Pond and Byram Township in Sussex County provide cartographic confirmation of their early presence on the landscape. I extracted the history you will encounter on the following pages from a much larger body of research of greater compass, geographically and chronologically, which I hope to publish in separate, digestible bits.

Place in History

Writing in 1859, J. Peter Lesley, secretary of the American Iron Association, described the Highlands of northwestern New Jersey as a "belt of short, parallel, half disconnected, half confused mountains of nearly the oldest rocks we know."[1] But it was the vast mineral resources, particularly iron ore, embedded in these deeply eroded hills that commanded his professional interest. In 1856, State Geologist William Kitchell enumerated and described upward of eighty iron mines, located in the counties of Sussex, Passaic, Morris and Warren, all within an area of 360 square miles. He recalled how, "in early days," these Highlands mines "furnished a very large portion of the ore manufactured into iron in this country, yet they have been excavated to a very limited extent, many of them containing immense bodies of ore above water-level, which may be economically extracted without the employment of expensive machinery."[2] In 1864, his successor, Professor George H. Cook, further narrowed the field, noting, "These beds are not uniformly distributed through the gneiss rocks [of the Highlands]; in some parts they abound, while in others no mines of value have ever been found. The most productive mines as yet worked have been in the central parts of the range, but as the demand for ore increases other mines are being sought for in the less promising districts."[3]

Andover Mine, situated on the western margin of the Central Highland Plateau in Andover Township, Sussex County, attracted particular attention and admiration. J. Peter Lesley spoke glowingly of the Andover ore body in 1859, saying, "Altogether it is by far the most interesting and perhaps

ANDOVER IRON MINE, SUSSEX CO.

Andover Iron Mine, Sussex Co. *From* Second Annual Report on the Geological Survey of the State of New Jersey, for the Year 1855.

the most important vein of the mining region."[4] Geology professor James T. Hodge, Lesley's longtime friend and colleague, favorably compared Andover with recently discovered iron ore deposits in the Marquette Range of Michigan's Upper Peninsula. Writing in 1853, he noted, "Some of the Andover ore of New Jersey cannot be distinguished from the choicest of the Lake Superior ores; and if made into bar iron direct, with the same care as were the samples for trial prepared from this ore, there is no question but it would exhibit the same remarkable strength; the pig iron manufactured from it, though made with anthracite, possesses the strength of the best charcoal iron."[5] Due to its unique chemical composition and accessibility, Lesley noted the international esteem that Andover iron earned in the eighteenth century, reminding his readers, "This ore was worked for steel, for the manufacture of which it proved well adapted. The bar iron made from it too bore a high reputation for toughness."[6] So where does the largely untold story of Andover iron begin?

Discovery

Thomas Woolverton, a shopkeeper in Bethlehem Township, Hunterdon County, and his wife, Mary Pettit, settled on 312 acres at Huntsville, Sussex County, in 1750. Here he erected a log grist- and sawmill at the confluence of the upper branches of Pequest Creek. On November 22, 1752, he was appointed justice of the peace for the upper parts of Morris County. Seven months later, in response to swelling numbers of inhabitants, the territory lying north and west of the Musconetcong River formed the new County of Sussex. Woolverton was commissioned justice of the peace and Sussex County's first tax collector. On December 2, 1755, the governor ordered the courts for the fledgling county to be held at Woolverton's house until such time as a courthouse could be erected. Woolverton's tavern thus became the first county seat within the present bounds of Sussex County, holding this honor until a courthouse was erected on its present site in Newton in 1762. The outbreak of the French and Indian War interrupted the first court sessions as atrocities in the neighborhood made it unwise to leave homes defenseless; jurors were dismissed without being sworn.

On the south bank of Pequest Creek at Huntsville, opposite Woolverton's stone house, stands a stone smithy and wheelwright's shop. This is said to occupy the site of a bloomery, which Major Woolverton operated between 1757 and his death in 1760. Known as Bango Forge—supposedly from the incessant banging of its trip hammer—it deserves honorable mention in the history of American metallurgy as the first hearth to work up iron ore from the famed Andover mine.

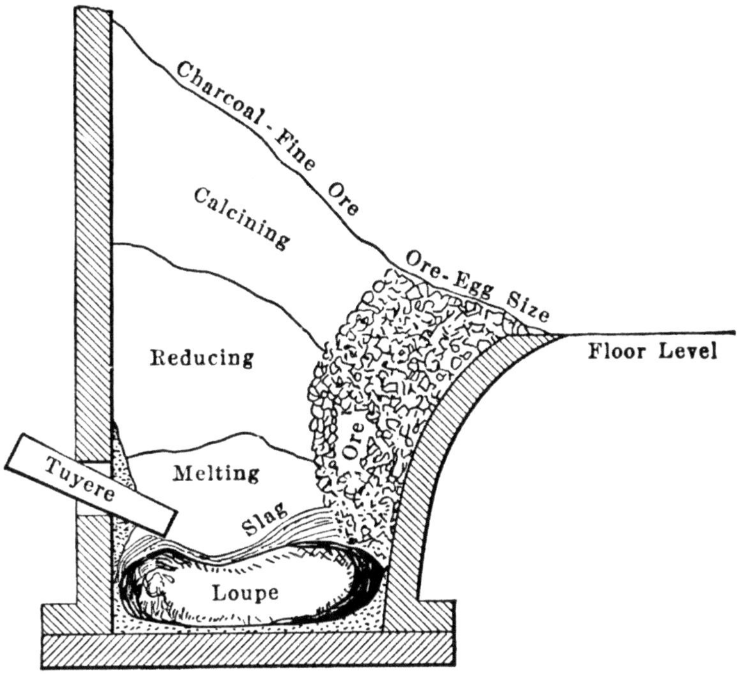

Fig. 8.—Catalan forge.

Diagram of Catalan forge used to illustrate direct wrought-iron processes. *From Richard Moldenke,* The Principles of Iron Founding *(1917).*

Today the Andover Mine Bank is but a waterlogged crevice in the wooded ridge east of Limecrest Road, nearly opposite Andover-Aeroflex Airport. It is reputed to have contained the most extensive hematite-magnetite deposit ever discovered in New Jersey and as high grade as any in America.[7] This great ferrous pod was imbedded within the bounds of William Penn's Tract No. 35, surveyed on May 23, 1714, encompassing 1,313 acres at the head of a branch of Pequest Creek. Penn's sons and heirs, Thomas and Richard, advertised this property in March 1755.[8] Whether Woolverton overheard discussion of its mineral treasure in his tavern or stumbled across the ore body during his own investigation of the neighborhood is a matter of pointless conjecture. With hardwood for charcoal in abundance and with the steady pulse of the Pequest for power, Woolverton established a frontier bloomery.

Replicating the primitive Catalan forge, a *bloomery* produced wrought-iron bars directly from rich magnetic ore without a blast furnace, using instead "a hearth of stonework, from six to ten feet square, in which there is a fireplace from twenty to thirty-six inches square, and from fifteen to thirty inches deep, lined with cast-iron plates."[9] A charge of pulverized ore, ground to the consistency of coarse sand, was converted in about two hours at about eight hundred degrees Fahrenheit into a pasty iron sponge ball, twelve to fifteen inches in diameter, weighing about three hundred pounds.[10] This *loupe* or *fining* was set on the floor and beaten with wooden clubs to break up any remaining slag. Reheated to welding temperature, it was shaped under a water-powered hammer into a rough square or *bloom*, driving out most impurities. The bloomer chopped this rough iron slab into several manageable pieces, which he reheated and flattened into bars. Hammering continued for sixteen grueling hours per day, outputting four blooms, each weighing from seventy-five to one hundred pounds. A ton of bar iron could be manufactured in a six-day workweek.[11] Nineteenth-century estimates suggest a bloomery, working night and day for a week, furnished "from four to seven tons of blooms."[12] Considering everything that could and did go wrong, a bloomery with two hearths produced about seventy tons of bar iron annually.

Bloomery iron could be forged and fashioned into such everyday necessities as nails, shoes for draught animals, tire iron and farm tools, but breaks could not be easily welded. Although estimates varied, bloomeries wasted prodigious quantities of fuel and ore. Some calculated a ton of blooms consumed from 3,500 to 5,000 pounds of charcoal and from one and a half to one and three-quarters tons of pulverized ore in its manufacture.[13] State Geologist William Kitchell estimated that two and a half tons of ore and five hundred bushels of charcoal were used to make a single ton of bar iron.[14] Skilled labor on the frontier was always expensive, and Bango Forge steadily emptied its owner's purse. Neighbors thought the monotonous resounding call of its trip hammer wailed over and over again, "Come penny, go pound."[15]

British imperial economic policy encouraged such enterprise. To attract a cheap and plentiful supply, Parliament removed all customs duties on pig iron imported into Great Britain from the American colonies in 1750. To limit competition in finished wares, however, the act prohibited colonists from building any rolling or slitting mills or plating forges that used a tilt hammer or any steel furnaces.[16] Iron mills erected before passage of the act were allowed to continue. As greater encouragement, Parliament permitted bar iron to be imported duty free in 1757. The French and Indian War

Fig. 74.

Blomary fire.

Bloomery. *From Frederick Overman,* The Manufacture of Iron, in All Its Various Branches *(1854).*

stimulated demand and inflated prices, luring English and American capitalists to invest in charcoal-iron plantations of vast extent on a frontier endowed with seemingly unlimited resources of water, ore and wood.[17] These investors imported German and English workers and shipped their output cheaply as ballast. Capacity expanded rapidly.

As a Bethlehem merchant, Thomas Woolverton lived and traded with employees of the Union Ironworks on the South Branch of the Raritan

Above: Hammering the bloom under a trip hammer. *An Iron Forge*, Library of Congress, Prints and Photographs Division, Washington, D.C.

Opposite, top: Oil painting of William Allen by Robert Feke, from life, 1746. *Courtesy Independence National Historical Park.*

Opposite, bottom: Joseph Turner by John Wollaston, circa 1752. *Courtesy Chester County Historical Society, West Chester, Pennsylvania.*

River, which two Philadelphia merchants, William Allen and Joseph Turner, established on 3,100 acres acquired from the West Jersey Society in 1742.[18] A rolling-and-slitting mill for producing iron plates and nail rods was added to their operations before 1750. A sample of the newly discovered ore from the upper reaches of the Pequest Creek (or perhaps some bar iron wrought at Bango Forge) attracted the attention of John Hackett at the Union Ironworks. His principal supply of ore at High Bridge was riddled with sulfur, which, even in comparatively small quantities, caused structural weakness. In contrast, the lode of red hematite at Andover was commingled with magnetite and rich in manganese. If combined with lime, manganese

acts as a flux to free the iron of impurities.[19] Manganese also acts as a hardening agent—an important virtue in the age of sand casting—and increases the handling life of molten metal as it was being poured into molds. Most importantly, a manganiferrous iron was singularly suited for conversion into wrought iron and steel. Soon after its introduction, Andover iron set the standard by which other ironmakers compared their metal. Even allowing for extravagant claims in advertising, there is sufficient contemporary testimony to conclude the ore was "esteemed of the best quality of any in America," proven "from experiments, made both in England and America, to be proper for every use to which iron can be converted, and equal to the Swedish for making steel."[20]

John Hackett relayed word of his discovery to his employers. Judge William Allen, son and namesake of a Scots-Irish trader, had studied law in England.[21] On his return, he married well and climbed easily to the pinnacle of Philadelphia society, successively serving as customs collector, provincial

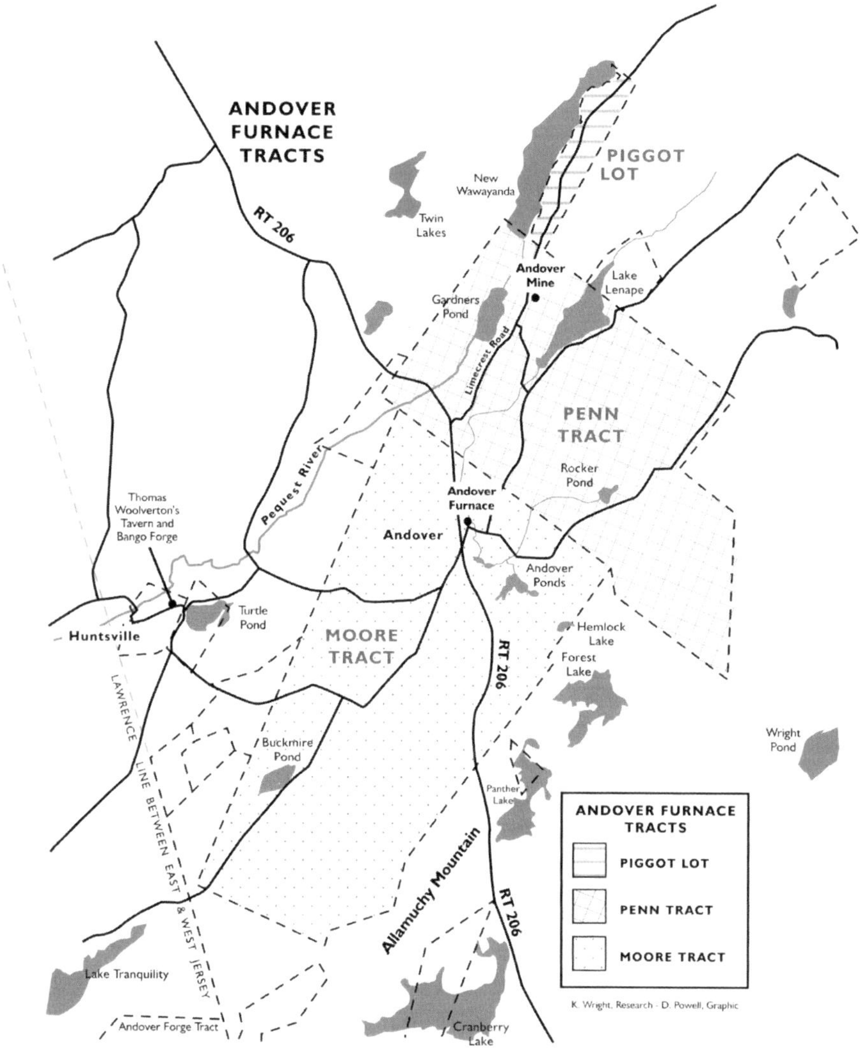

Map of Andover Furnace Tracts. The Andover Iron Company acquired 4,054 acres for the operation of its furnace. *Courtesy Deborah Powell.*

assemblyman, city councilman and then mayor. His daughter Ann married John Penn, Pennsylvania's last provincial governor. Well connected, William Allen speculated extensively in frontier real estate, especially in the Lehigh Valley, where Allentown bears his name. In 1725, William Allen formed a commercial partnership with a sea captain and boat builder from Andover, England, named Joseph Turner.[22] They conducted an international trade in sugar, molasses, rum, enslaved persons, iron and horses. Allen and Turner purchased a share in Durham Furnace in 1727 and erected the Union Ironworks in 1742. William Allen was appointed supreme justice of Pennsylvania in 1750, a position of highest honor, prestige and influence.

John Hackett was born in Ireland in 1728. Whereas Judge Allen and Captain Turner traipsed the Philadelphia wharves or graced the city's mansions and coffeehouses, John Hackett faced the lawless frontier. Squatters surrounding the Union Ironworks claimed title to their homesteads by purchase from the Indians and violently contested Allen and Turner's acquisition of their lands from the West Jersey Society. In 1748, they assaulted John Hackett and William Bird, manager of the Union Ironworks, with clubs, as the two men felled timber to build a mill.[23] Ringleaders of the so-called Club Men were arrested and escorted to jail in Trenton, but the trouble did not end there. In 1754, one hundred or more German squatters, angered by extensive deforestation, threatened to pull down the Union Furnace. Hackett calmly offered the rioters a "Dram," buying time for his employees to arm themselves and march to his rescue. In the ensuing brawl, someone punched Hackett in the stomach before the squatters retreated.

A year later, John Hackett married Elizabeth Reading, daughter of surveyor John Reading Jr. of Amwell.[24] A tombstone outside the Bethlehem Presbyterian Church marks the burial of Joseph Turner Hackett, son of John and Elizabeth Hackett, who died August 18, 1761, aged twenty-two months. John Hackett named the Andover Iron Company to honor Turner's English birthplace.

Andover Furnace

Organized in 1758, the Andover Iron Company comprised William Allen and Joseph Turner, who each owned a five-sixteenths interest; the well-connected, wealthy landowner Lynford Lardner, who owned a quarter share; and junior partner John Hackett, who owned an eighth share.[25] The new company began acquiring proprietary rights to unclaimed lands in the mountainous district of Sussex County in April 1758, eventually paying £1,120 for rights to 4,287 acres. On August 11, 1759, Thomas and Richard Penn, acting through attorneys Richard Peters and Lynford Lardner, sold Penn Tract No. 35, containing 1,313 acres, to John Hackett for £500. This property encompassed the Andover ore deposit. The present Borough of Andover occupies the northeastern third of an adjacent tract of 1,666 acres, which John Hackett bought from William and Willianna Moore, of Chester County, Pennsylvania, for £574 in 1759.[26] Andover Furnace and Mills were sited where a small tributary of Pequest Creek tumbles through a gravelly dell among glacial kames, offering the nearest suitable water power to the new mine. Hackett also acquired 15.5 acres at the outlet of Panther Pond for an additional reservoir to power the ironworks during any protracted drought. In 1760, Colonel Hackett acquired the 82-acre Piggot Lot, including the southeast shore of Piggot's Pond, now Lake Aeroflex in Kittatinny Valley State Park.[27] He also acquired 200 acres surrounding Hall's Pond, now Lake Iliff.[28]

Limecrest Road (Route 669) was laid out, 4.8 miles in length, in 1761, departing what is now Route 206 "at an old Lime Kiln" about 528 feet north of the "Andover House."[29] The road ran past "a Large Rock next

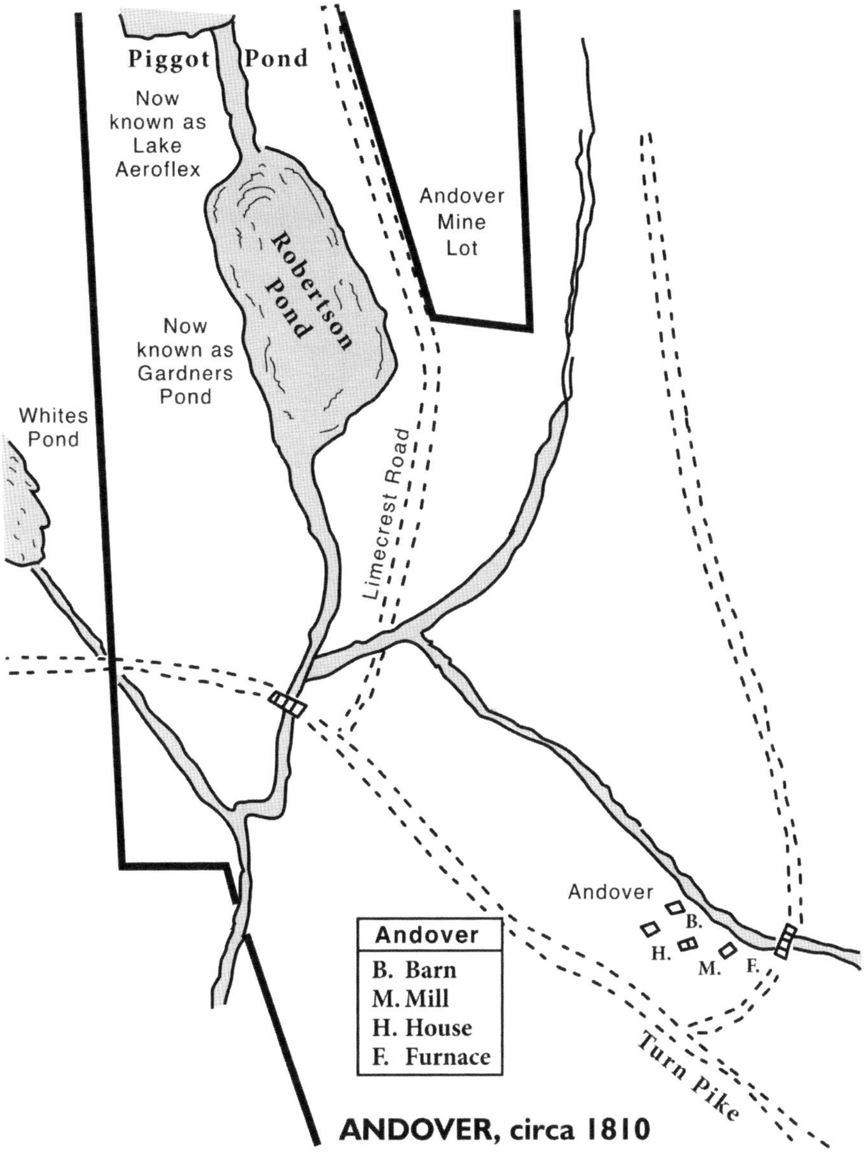

Map of Andover Furnace Farm and Andover Mine Lot, based on 1810 map of agent Frederick Bartles, shows location of iron company buildings in Andover. *Courtesy Deborah Powell.*

to Andover Mine Hill," continuing by "Doctor Piggot's old Stockyard," located somewhere in the vicinity of Long Pond Farm. It ran "to a Black Oak standing in a certain Road that leads from Justice Joseph Hall to

Fig. 51.

Vertical section of a blast furnace designed for charcoal.

Vertical cross section of charcoal blast furnace. *From Frederick Overman,* The Manufacture of Iron, in All Its Various Branches *(1854).*

Henry Simpson's near a sunken pond." This "certain Road" is probably Mulford Road and the so-called sunken pond is Howells Pond. Lake Iliff was originally named Hall's Pond for Joseph Hall.

In March 1760, John Hackett purchased 663 acres of woodland extending from the feeder stream of Lake Lenape eastward to include what is now

Perona Farms and Lake, then southwest to the borders of Valentine Pond on the Andover-Mohawk Road (Route 613).[30] In October 1761, Hackett acquired 119.5 acres on the east side of Panther's Hill, bordering the Great Road (Route 206) and Cranberry Lake. According to corrected surveys, the Andover Iron Company ultimately acquired 4,054 acres for the operation of its new furnace.

The eighteenth-century blast furnace was a truncated stone pyramid, approximately twenty-five to thirty feet square at the base, standing twenty-five to thirty-five feet tall, with a chimney built on top.[31] Molten iron and slag collected in a cylindrical reservoir centered at its base called the crucible. The interior walls above the crucible sloped outward to a diameter of about eight feet, forming a deep basin called the bosch.[32] This outward flare slowed the velocity of ascending gases to enhance their reductive action, and it served as an internal buttress to support the weight of the three-and-a-half-ton charge. To reduce friction on the descending stock, the shaft, rising above the bosch, took the shape of an inverted cone, gradually contracting until the charging-hole at the top was no more than two feet wide. Slate was used for a furnace lining at Andover. A newspaper advertisement for land, which Jonathan Hampton posted on August 23, 1760, mentions "Mr. Hackett's new Furnace."[33] On March 25, 1761, William Allen wrote John Griffiths, relating, "Our manager, Mr. Hackett, says the new Iron we make yields so well that less than 26 cwt. Of Pigs will make a Ton of Bar…Since we have wrote you, we have finished our small Blast and from the first Day of September [1760] to the first of December our Furnace produced 332 Tons of Pig metal, which was beyond our most sanguine expectations."[34]

The furnace was "blown in" by filling it with cordwood and charcoal and then lighting it from the top. A steady air blast produced a hot fire, which burned down after several days to the level of the pipes supplying the air blast. Chargers reloaded the stack with fuel, allowing the fire to work its way back up, heating the inwalls and expelling moisture. Measured basket loads of ore, charcoal, limestone and clay were fed into the furnace's throat in alternating layers, so as not to crush the charcoal. Clay acted as a lubricant. Very little ore and flux were added at first, the proportion gradually increasing until the full burden was reached after a day or two.[35] A normal charge of 230 pounds of charcoal (about 15 basket loads), 500 pounds of ore, 50 pounds of limestone and 20 pounds of clay might be added every half hour. The typical Highlands furnace consumed about 36 tons of ore, 2.75 tons of limestone and almost 6,000 bushels of charcoal weekly in the production of 20 tons of pig iron.[36] Describing procedures at Wawayanda

Above: Chargers load furnace with baskets of charcoal through the bridge house. *Charles Gillespie, editor.* A Diderot Pictorial Encyclopedia of Trades and Industry, 485 Plates Selected from "L'Encyclopédie" of Denis Diderot. *Courtesy Dover Publications.*

Opposite: Hopewell Furnace National Historic Site, National Park Service: (1) coal house; (2) bridge; (3) bridge house; (4) furnace; (5) flume; and (6) wheel house. *Photograph by Deborah Powell.*

Furnace in 1856, William Kitchell reported each charge consisted of eleven baskets of charcoal (equivalent to 22 bushels), 525 pounds of ore (reduced to the size of hickory nuts) and from 30 to 50 pounds of limestone.[37] Every twenty-four hours, the furnace received fifty to sixty charges, yielding 7 to 8 tons of iron.[38] Andover Furnace produced an average of 18 tons weekly, though at times, 20 to 22 tons per week were obtained.[39]

Since iron comes from the earth as an oxide, a chemical reaction in the blast furnace released metallic iron from its ore.[40] Through an arched alcove about ten feet wide on one side of the furnace stack, a mechanical apparatus blew air through pipes called tuyeres. Continuous gusts of oxygen burned charcoal to generate both heat and the carbon monoxide needed to extract a sponge of relatively pure iron. This iron sponge liquefied at about 2,400 degrees Fahrenheit, losing most of the remaining oxygen as the iron rapidly absorbed carbon. Metalloids such as phosphorus and silicon combined with the limestone flux to be drawn off as molten slag.[41] On occasion, insufficient heat or excessive moisture rendered the charge a pasty mess, choking the furnace by forming a ring on the sides or a *scaffold* across the interior or simply a large indigestible mass called a *skull*. *Pillaring* occurred when the air blast was not sufficiently penetrative and a column of cold stock clogged the middle of the hearth. Uneven burning produced hot spots in the furnace walls and a possible breakout.[42]

Hopewell Furnace National Historic Site, National Park Service: (1) shed for unloading charcoal; (2) coal house; (3) bridge; (4) bridge house; (5) furnace; (6) casting and molding house. *Photograph by Deborah Powell.*

Remnants of the Andover Furnace Pond Dam. *Photograph by Deborah Powell.*

During smelting, pellets of molten iron, containing between 3.0 and 4.5 percent carbon, trickled into the crucible. As the metal cooled and hardened, carbon remained either mechanically mixed with the iron as crystallized carbon (graphite) or dissolved into a solid solution known as combined carbon.[43] Combined carbon made cast iron brittle. The relatively high temperatures attained in the furnace reduced not only nearly all the iron, manganese and phosphorus present in the ore but also some of the silicon, sulfur and other elements. These entered the finished product in varying concentrations and decidedly affected its character. Silicon determined whether carbon segregated as graphite or was retained in combined solution, properties of particular importance when casting.[44] Low-silicon pig iron has a higher content of total carbon. Manganese increased the combined carbon content and aided desulfurization.[45] Sulfur was prone to segregate and thus weaken the metal.[46] Phosphorus caused iron to solidify in granules, making it hard and brittle.[47] Owing to its crystalline structure and relatively high carbon content, pig iron was too hard and brittle to be rolled or forged at any temperature.[48]

Bridge and casting houses, a molding house and a wheel house surrounded Andover Furnace, providing workshops or sheltering the machinery.[49] Stockpiled ingredients were measured in baskets and carried in wheelbarrows through the bridge house, a covered passageway leading from the coal house on an adjacent hill to the charging hole atop the furnace. Since cool, dry air improved fuel efficiency, smelting was commonly a winter operation. Accordingly, the wheel house was attached at the tuyeres-arch to prevent snow and ice from arresting the motion of the water wheel and bellows. To create a twenty-eight-foot fall of water, the furnace was built in 1760 about a fifth of a mile below the Furnace Pond Dam.[50] A system of excavated races, wooden troughs and pipes, made of hollowed tree trunks, conducted water from the Furnace Pond to the water wheel, allowing unobstructed passage of roads approaching the furnace. A cascading stream of water drove an overshot wheel, twenty-three feet in diameter. Four cams on a rotating shaft repetitiously struck an armature fastened to the upper plate of the bellows, thus compressing a mechanical lung and forcing air through a blowing pipe into the smelting chamber. A long armature fitted with a counterweight re-inflated the bellows.

In 1761, William Allen informed English ironmaster John Griffiths that a German introduced wooden bellows, which not only proved superior to those made of leather but also lasted twenty years. Consequently, "no other than Wooden Bellows are now used among us either at our Furnaces or Forges."[51]

Fig. 181.

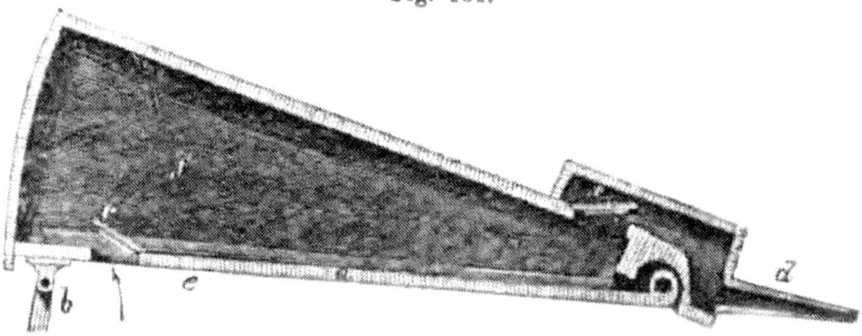

Swedish bellows.

Fig. 182.

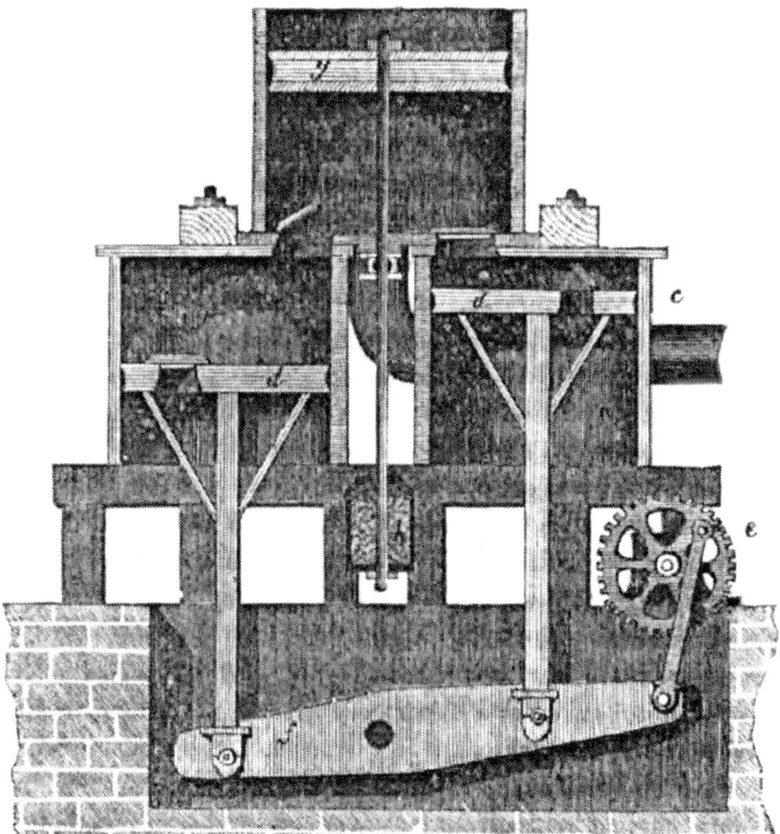

Wooden cylinder bellows.

Most likely, Allen was referring to a kind of blast machine, called Widholm's bellows, extensively used in Sweden, Russia, Germany and France but rarely in England or in British North America.[52] This apparatus looked exactly like a blacksmith's bellows, except that it was fashioned entirely of wood, measuring six or seven feet in length and thirty inches in width. An iron rod connected with a crank on the water wheel pumped the hinged wooden base of the bellows at ten to twelve strokes per minute. Usually, a pair of bellows worked alternately to supply a continuous blast of air, one exhaling as the other inhaled. Otherwise, William Allen may refer to wooden blowing tubs with leather valves.[53] Such reciprocating blowing engines consisted of two thick-walled wooden tubs, fitted one inside another, the bottom one being fixed while the other worked as a piston to compress air and blow it into a leather bag.[54] A coating of graphite or lead ore, applied between the tubs, served as a lubricant. Pressure was only about three-fourths of a pound per square inch with the piston making fifteen strokes per minute.[55]

Molten iron flowed from the crucible beneath a *tymp-stone*, concentrating in a pool on the hearth behind the *dam-stone*. In the attached casting house, iron was tapped through a work arch, fourteen feet at its widest, about every twelve hours by drilling out a clay plug at the base of the dam-stone.[56] Depending on capacity, a blast furnace could produce up to five or six tons daily.[57] The hot liquid iron ran down a long feeder trench and into numerous lateral branches dug in a sand bed. This pattern of furrows resembled a sow nursing piglets, and the hardened slabs were consequently called "pig iron." As slag is lighter than iron, a clay-washed cast-iron plate was set in the main feeder trench to skim off any slag coming out the iron notch.[58] Slackening the blast, clay balls were pressed into the iron notch to seal it again. Slag was flushed out slightly higher through a cinder-notch, using a pointed iron bar. Another bar with an enlarged end was then held against the hole until the cinder cooled against it, thus closing the tap. Andover Furnace probably commenced operations on October 29, 1761, producing at least fifteen tons of pig metal by November 25, for a daily yield of half a ton.[59]

Opposite, top: Wooden bellows, known as Widholm's bellows. *From Frederick Overman,* The Manufacture of Iron, in All Its Various Branches *(1854).*

Opposite, bottom: Wooden cylinder or tub bellows. *From Frederick Overman,* The Manufacture of Iron, in All Its Various Branches *(1854).*

While touring American ironworks, including Andover, in 1783, Swedish ironmaster Samuel Gustaf Hermelin observed how "a great quantity of castings are manufactured, all in sand forms, such as iron fireplaces and small stoves, pots, pans, flat irons, wheel bushings, etc., and during the [Revolutionary] war 3, 4 to 6 pound cannons, cannon balls and bombs were manufactured."[60] For fire backs, liquid metal from the furnace filled an impression left by a carved wooden pattern in the sand floor of the casting room. Most utensils, however, were hand cast in prepared molds called *matrices*. Damp, greasy sand was tamped around a wooden pattern within a *molding box*. Since metal contracts as it cools, the patterns were made slightly larger than the desired casting. With two-part molds, opposite sides of the pattern were set in matching frames. When the sand dried, these molds were fitted together to form a void identical to the shape of the object to be cast. Clamped together, the two halves were set in a press under heavy weight. The casters pierced the sand matrix with a wire to open vents for gases.[61]

As it was risky to tap the furnace for small quantities of iron, hand casting was common. Molders ladled molten iron from the fore-hearth into a feeder tube attached to the molding box, which fed small cylinders set in the sand to serve as conduits, filling the void in the mold from the bottom.[62] Cast wares were made in the molding house, adjacent to the casting house, allowing easy access to the hearth.

Above: Molders at work. *Charles Gillespie, editor.* A Diderot Pictorial Encyclopedia of Trades and Industry, 485 Plates Selected from "L'Encyclopédie" of Denis Diderot. *Courtesy Dover Publications.*

Opposite: Molders set patterns in sand. *Charles Gillespie, editor.* A Diderot Pictorial Encyclopedia of Trades and Industry, 485 Plates Selected from "L'Encyclopédie" of Denis Diderot. *Courtesy Dover Publications.*

fig. 4
fig. 6

Each week Andover Furnace consumed on average the charcoal made from six acres of forest.[63] The cost of fuel per ton of iron manufactured amounted to double the combined expenses for ore, limestone, foundry and clerical wages. Charcoal for several months' operation was stored in a coal house, which stood on the furnace bank. At birth, Andover Furnace was well positioned to satisfy its appetite for fuel, standing amid a virgin hardwood forest of oak, chestnut and hickory. In advance of agricultural settlement, such extensive tracts of timberland were cheaply acquired. This advantage stirred undue optimism in William Allen, who wrote in March 1761, "Water & Wood we can never want, indeed we want for nothing but good workmen."[64]

Charcoal is nearly pure carbon, obtained when wood is charred to remove moisture, thereby eliminating about 74 percent of its original volume.[65] For centuries, it was manufactured in dome-shaped piles, covered with turf or dirt, with small vents to control air intake and to allow gases to escape.[66] A master collier supervised fifteen to thirty woodcutters during the autumn cutting season, October through January, when the sap was down. Saplings between twelve and twenty-four inches in diameter were preferred to old-growth timber. An industrious woodcutter could average three cords per day.[67] In 1783, woodcutters received three shillings per cord for cutting and another three to four shillings per cord for sledding wood to the firing place.[68] Hickory, chestnut and white, red and black oaks were preferred.

To prepare a coaling pit, charcoal burners raked a level piece of dry ground in a sheltered hollow, forty to fifty feet in diameter, choosing as accessible a location as possible.[69] Since gusting winds would ignite the smoldering heap, colliers usually worked during the settled weather of summer and fall. At least five hundred cords of wood were needed for a single coaling to be profitable.[70] The collier set from one to three poles of green wood, called a *fagan*, to form a draft hole in the center of the firing place, piling a triangular chimney of logs against it. Next, four tiers of cordwood were stacked in rings radiating outward from this central chimney, placing heavy billets of wood on the bottom and lighter pieces of lapwood (or logging residues) on top.

Opposite, top: Illustration of Andover Furnace. *From "New Jersey Invites You—No. 17 by G.A. Bradshaw,"* Somerset Messenger and Gazette, *April 26, 1940.*

Opposite, bottom: Constructing charcoal mounds. *Charles Gillespie, editor.* A Diderot Pictorial Encyclopedia of Trades and Industry, 485 Plates Selected from "L'Encyclopédie" of Denis Diderot. *Courtesy Dover Publications.*

The first billets were set on end, supported by an encircling dirt mound, with successive layers sloping toward the center, building a tightly packed dome of wood. The collier stuffed wood chips into crevices and laid a smudge blanket of leaves and dust, several inches thick, to prevent combustion. A standing kiln of twenty-five to thirty cords, measuring ten to fourteen feet in diameter, took seven days to coal.[71]

The collier removed the fagan to open a flue in the center chimney, which he packed with kindling. He then dropped in hot coals. Complete charring and careful cooling took seven to nine days, during which time the collier used log plugs to open and close air vents drilled into the sides of the mound to warm cold spots or stifle any outbreak of fire. Colliers resided in shanties beside their kilns, keeping constant vigil to ensure even charring throughout the heap as intense heat expelled sap and moisture as vapors. As hot, dry charcoal absorbs oxygen rapidly, the smoldering heap might catch fire before it cooled sufficiently to handle. To prevent this, the collier raked out only as much charcoal from his kiln as a coal wagon could carry in a single load. As late as 1793, a survey for land situated on Andover Mountain, about two miles northeast of Andover Furnace, mentions "an old coal road that crosses the swamp" between James Smith and Richard McPeek.[72]

Depending on the quality of the timber, an acre of woodland yielded between sixteen and thirty cords of wood.[73] By some estimates, a cord of wood yielded 25 to 45 bushels of charcoal, depending on the species used, the collier's skill and the standard of measurement.[74] One eighteenth-century engineer estimated two to two and a half cords of wood yielded about 100 bushels of charcoal (or 40 to 50 bushels per cord).[75] A large kiln produced as much as 1,350 bushels—bushels varied from 2,480 to 2,748 cubic inches in volume and from 18.0 to 22.5 pounds in weight.[76]

The Andover Iron Company used four coal wagons to transport charcoal to the coal house. A 1772 inventory mentions "1 Old frame of a Coal body without boards," suggesting these coal wagons had removable bottom boards to dump their loads on the ground without shoveling.[77] As precaution against fire, charcoal brought to the coal house was first spread on the ground for further cooling and careful inspection. A shed extension of the roof protected the cooling charcoal from precipitation. Although no trace of it is evident today, the coal house at Andover was likely built into the hillside above the furnace and required "a bridge to bring the coals in." Charcoal was dumped into the cavernous stone building from this entry ramp and removed from the opposite and downhill side of the building through doors leading into the bridge house. A "window in the roof" provided ventilation

Fig. 27.

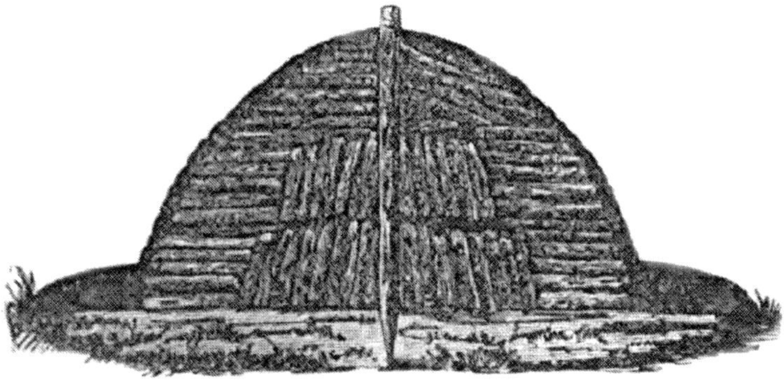

Section of a charcoal work—piling the wood.

Section of a charcoal work—piling the wood. *From Frederick Overman,* The Manufacture of Iron, in All Its Various Branches *(1854).*

Fig. 26.

Making charcoal in heaps.

Making charcoal in heaps. *From Frederick Overman,* The Manufacture of Iron, in All Its Various Branches *(1854).*

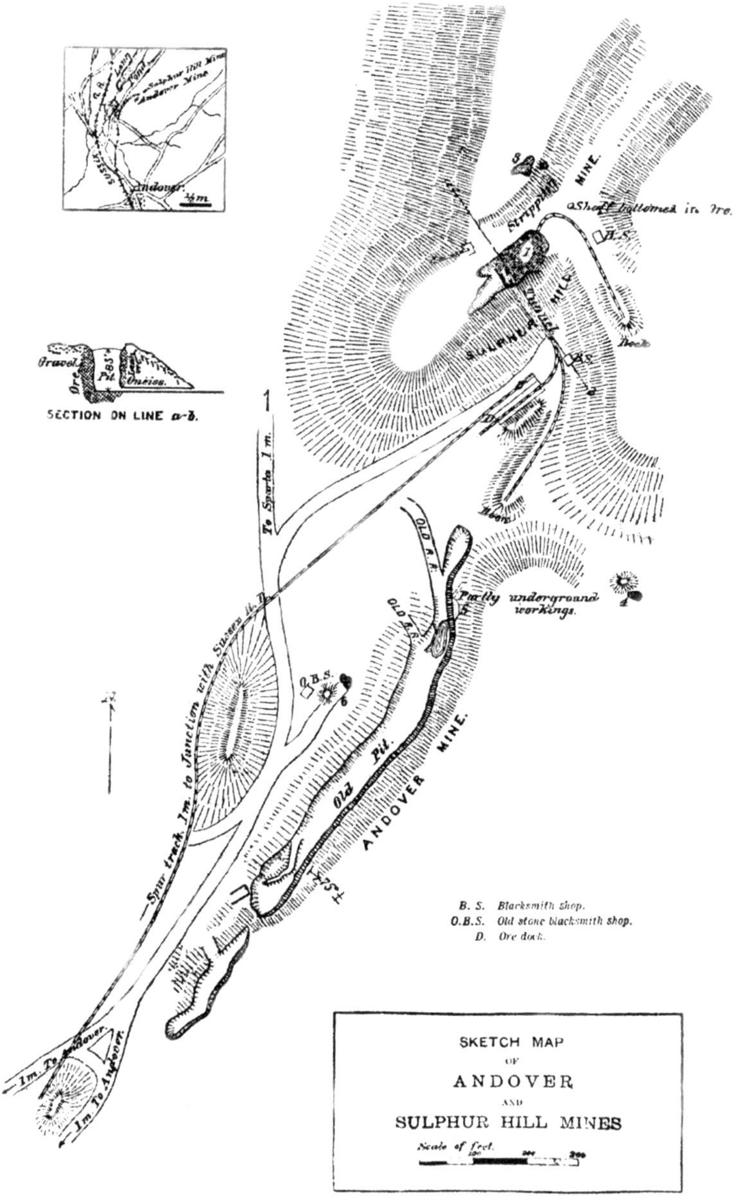

Fig. 4.

(From 10th Census Reports, Vol. XV, p. 152.)

The colonial mine is the "Old Pit" in this "Sketch Map of Andover and Sulphur Hill Mines." *From William S. Bayley,* Iron Mines and Mining in New Jersey *(1910).*

and illumination. An advertisement for lease of the Andover Ironworks, published in October 1772, mentions "a large Coal house in which there is at least 7 Weeks' Stock of Coal for the next blast."[78] As the furnace consumed between fifty-six and sixty loads (at one hundred bushels per load) of charcoal weekly to produce between eighteen and twenty-two tons of pig iron, seven weeks' stock amounted to about forty-two thousand bushels of the black, glistening fuel.

The "inexhaustible Body of Ore" at Andover was actually an exposed surface pod, embedded in the side of a ridge, where ore could be "raised at the easy Expense of 2*s.* per Ton" without deep shafts, whims for hoisting ore or pumps to drain water.[79] Hand tools inventoried in 1772 comprised forty-one borers at the mine hill, three crowbars, three stamp bars, six picks, two old shovels, four hammers, two rakes, three scrapers, two old sledges, three old wheelbarrows, four needles and three worms.[80] In 1783, the Andover Iron Company paid miners two and a half shillings per ton.[81] Washing rid the ore of earthy matter. Roasting it with charcoal in an open pile also removed some impurities and improved its quality for smelting.[82]

The Andover Iron Company constructed a stone gristmill a few yards northwest of its furnace, laying one hundred feet of hollow tree trunks underground to divert water from the furnace headrace to the mill. This primitive pipe, measuring two feet square, supplied water to a wooden trough called the penstock, where sluice gates regulated the flow to a high-breast water wheel, twenty feet in diameter.[83] The mill was internally divided into two chambers: a wheel house to the east sheltered the water wheel, and a mill house to the west housed the grinding mechanism. The water wheel turned a large cogwheel, fixed on the opposite end of the main axle. Wooden teeth, protruding from near its outer circumference, ran between wooden posts of a small lantern gear at about a five-to-one ratio, transmitting power to a vertical shaft. On the upper story, a heavy grinding stone balanced on an iron rind atop this upright shaft. The runner stone worked against a stationary bed-stone, fixed in the floor of the grinding platform. The vertical shaft carrying the runner stone sat in a cup bearing fitted on a heavy wooden lever called the *bridge-tree*. One end of the bridge-tree was pinned into the upright posts of the crib or grinding platform. A vertical iron rod connected the free end of the bridge-tree to a smaller lever on the grinding floor, allowing the miller to raise or lower the runner stone and thereby adjust the fineness of his grind. To prevent vibrations from shaking the building, the crib stood independent of the mill.

Joseph Northrup converted the gristmill and wheel house of the Andover Iron Company to a store in 1816, adding a distillery in 1828. *Photo by Kevin Wright, 1981.*

The miller dumped grain into a wooden hopper above the grinding tubs. Kernels dropped through an opening at its base into a small, angled wooden chute. During grinding operations, this chute constantly bounced against the ribs of a small vertical post, called the *damsel*, fitted atop the rind supporting the runner stone. The "chattering damsel" repetitiously shook handfuls of grain from the chute into the eye of the runner stone as it turned. The miller regulated the flow of grain by raising or lowering the feed end of this chute. Working its way from the center to the outer edge of the millstones, grain was ground against a pinwheel pattern of furrows cut into the grinding surfaces. Where the furrows intersected near the eye of the stone, their shearing edges snipped away the kernel's tough hull. Near the outer edge, natural pores and irregularities in the rough field between the furrows finely ground the starch into flour. The pinwheel design of furrows kept centrifugal force from speeding grain across the grinding surface too rapidly. These small channels also dissipated heat. After processing approximately twenty tons of grain, the miller dressed or re-sharpened the furrow edges with a pick hammer. Calipers attached to a crane beside the grinding tubs could lift the runner stone and turn it over for dressing. Once set, the miller listened

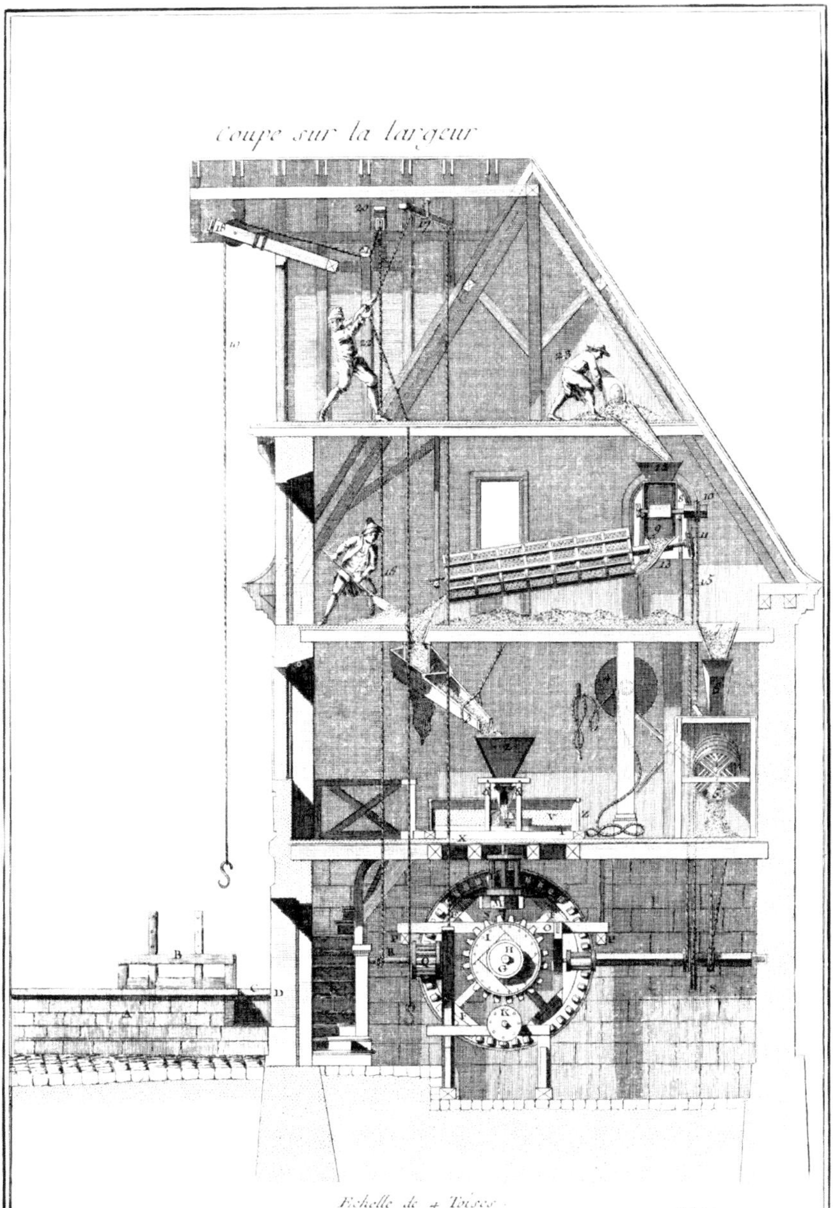

Cross section of gristmill. *Charles Gillespie, editor.* A Diderot Pictorial Encyclopedia of Trades and Industry, 485 Plates Selected from "L'Encyclopédie" of Denis Diderot. *Courtesy Dover Publications.*

to the "singing" of the millstones to determine whether the furrows needed dressing, whether the speed of rotation needed adjustment or whether the flow of grain between the stones was properly regulated. Besides grinding meal and flour, the mill cracked grains for animal feed.

Leased to a private operator, the Andover Gristmill developed "a great Deal of Country Custom," whereby the miller kept a toll (traditionally one-seventh) of the grain that he ground.[84] A beam and scales weighed incoming grist. When inventoried in 1772, the Andover Mill stored 483 bushels of wheat, 110 bushels of rye and 103 bushels of corn.[85] The old gristmill and wheel house still stand, albeit much altered over a long and useful life.

A blacksmith shop operated near Andover Furnace, and contemporary inventories list two pairs of smith's bellows, two anvils and sets of blacksmith and wheelwright tools, indicating the operation of two hearths under its roof.[86] There was also a *buttress*, which was a wooden framework with leather straps used to hold a horse while its hooves were pared. Wheelwrighting and wagon repairs offered steady employment. In 1767, the Andover Iron Company utilized two wagons, each drawn by a team of six horses; two ox carts and five pairs of oxen; and four coal wagons.[87]

Andover still boasts "an elegant Stone Dwelling-house," built as the ironmaster's mansion.[88] Early inventories mention five pairs of andirons, four pairs of fireplace tongs and four ash shovels, indicating four fireplaces in the main house and another in the out-kitchen.[89] The manager and a dozen of his employees lived under the same roof. In 1767, its furnishings included eight feather beds and thirty old blankets.[90] In 1772, sleeping accommodations included ten feather beds, eight bedsteads and six trundle or "under Beds."[91] Produce from a fenced garden plot, covering an acre, supplied the out-kitchen. The springhouse provided household water and a cooler for dairy products. Normally, a head smelter and caster, a smelter's assistant, two mixers, an ore-roaster, three ore and limestone buckers, two drivers and one coal measurer attended the furnace during operations.[92] The company also employed an inspector and bookkeeper. Benjamin Cooper was the first ironmaster.[93]

Animal power was indispensible to the operation of a charcoal iron plantation. An inventory, dated June 8, 1772, mentions "1 Team of 6 horses with Wagon & gears drove by Walker," valued at £100, and "1 Team of 6 horses with Wagon & gears drove by Negro Mingo," valued at £90.[94] It also listed a brown horse (bought from John Pettit); a small bay horse; two large sorrel horses and a small sorrel horse; two old horses; a blind gray horse; an old roan horse; a black horse, blind in one eye; and five pairs of oxen. To shelter these animals, a large stone

Farrier's buttress for hoof paring. *Charles Gillespie, editor.* A Diderot Pictorial Encyclopedia of Trades and Industry, 485 Plates Selected from "L'Encyclopédie" of Denis Diderot. *Courtesy Dover Publications.*

Andover ironmaster's mansion overlooks Route 206. Later known as the Northrop House, it was renovated after a 1903 fire. *Photo by Deborah Powell, 2013.*

The Andover mule barn or "Fort," demolished in 1954 to make way for a firehouse. *Courtesy of artist and illustrator William Gibbs.*

bank barn, including stables and wagon house, stood on the east side of the road (Route 206), below the ironmaster's mansion. Some old-timers remembered it as the "old mule barn," no doubt from some later use since the Andover Iron Company did not use mules. Because of several vertical slits in its gable ends—ventilators mistaken for musket ports—some called it the "Fort." Sadly, it was demolished in 1954 to make way for the firehouse now standing on Route 206 in Andover Borough.

The company store stocked provisions for ironworkers and attracted trade from the surrounding countryside. Merchandise comprised dry goods, cured meats, tobacco and rum, a variety of fabrics, bed covers, shoes and boots, buttons, hardware and every description of domestic utensil.[95] Touring American ironworks in 1783, Samuel Gustaf Hermelin noted, "The workmen at most of the works obtain little or no cash payment, but rather payment in goods. At those works, where the men obtain the highest wages, provisions are also the most expensive. On flour, pork and other necessities, when selling at retail, some works add 25 percent to the prices paid, when the same are sold per bushel or per barrel. Also, approximately 50 percent is added to the price originally paid for linen, broadcloth, sugar, tea, coffee and rum, which 4 last-mentioned articles factory workers, as well as people in general, consume daily in their households."[96] William Kirby, a British army deserter employed at the Andover Ironworks, complained that "when they had worked six months, if they had anything coming, they may perhaps get a few rags to cover their nakedness at a very dear price, but as for money they will get none though they have ever so much need of it."[97] He went on to say, "The wood chopper piled his wood so as to cheat the collier. The collier put his coal into baskets in such a manner as to deceive the ironmaster; and the ironmaster, not to be outdone, sold his provisions to the men at an extortionate price."

The old stone storehouse of the Andover Iron Company burned down on February 15, 1876, according to a local correspondent, who reported, "One of the ancient buildings of Andover (belonging to J.S.W.) burned down.[98] It was a relic of 1761: a small building, and in the early days occupied for a schoolhouse, a store and a carriage barn. The roof appears to be of cedar put on with hand made wrought nails, and bears the marks of rain drops trickling down its slope for more than a century. Its fall has been expected for some time and it was unoccupied."[99]

The Andover Iron Company maintained "a number of Out-houses for Workmen."[100] Most were log houses, though a few may have been of frame or even stone construction. The company also operated two farms on the Furnace Tract. The first was described in 1770 as "a large Farm… on which about 60 Acres of Winter Grain is now growing, about 70 Acres of extraordinary good Meadow has already been made, and is now in good English Grass, and as much more may be cleared at a small Expense, it being already drained. The whole is under good Fence and the woodland adjoining affords excellent Range for a large Stock."[101] In 1767, the company's livestock included twenty-two horses, fifteen cows and five pairs of oxen. In 1782, John Jacob Faesch described this farm as containing about fifty acres of improved meadow, which needed "to be plowed at least three times & harrowed four times besides the Sowing and Timothy Seed, and one fourth part of the 50 Acres must be bogg'd, most of the inside ditches are filled up."[102] The fence needed three thousand rails and 250 posts to keep out cattle. This meadow was about one and a half miles from the furnace, but "the posts and Rails cannot be got nearer than one Mile." Faesch listed another farm of one hundred acres, needing two thousand rails and 250 posts for fencing. In May 1772, these farms had thirty-three acres of rye in the ground, two acres of potatoes, nine and a half acres of corn in the ground and five and a quarter acres of wheat in the ground. The company's farm tools included axes, shovels, wheelbarrows, brush scythes, dung forks, pitchforks, three plows with irons, a harrow and five harrow teeth.[103] Despite operating its own farms, the Andover Ironworks furnished "a good Market

Postcard view of log house near Waterloo, Sussex County, New Jersey, circa 1915.

for all Kinds of Country Produce," profiting farmers and craftsmen for miles around.[104] Farmers also hired on as teamsters and woodcutters in the fallow season.[105]

The Andover Saw Mill stood nearly a mile northeast of the furnace. A stone dam was built below the confluence of two small brooks, creating a millpond in a swampy basin.[106] In 1926, a dam near the same location formed Lake Lenape. The sawmill was probably no more than a wooden shed covering the machinery. Since soaking drew out the sap, timber, felled in winter and dragged on sleds, was rolled into the millpond for curing. Sawmills generally operated after a spring thaw when meltwater swelled mountain streams. A flume, fifteen feet long, eight feet high and five feet wide, conveyed water from the pond to a small impulse wheel.[107] This rapidly turning *flutter-wheel* drove the saw frame up and down by means of a crank armature and connecting link called the *pitman*. Each stroke of the reciprocating saw frame turned a ratchet gear (called the *wrag wheel*) that advanced the log carriage until the log was sawn along its entire length. To reverse the motion of the carriage, a lantern pinion was fitted to a vertical shaft turned by a *tub-wheel*. This lantern pinion engaged teeth on the wrag wheel and returned the log carriage to its starting position.

Andover Forge

Charcoal blast furnaces produced high-carbon cast iron, a brittle metal that cold bending or hard blows cracked.[108] To make it malleable, forgemen refined pig iron at relatively low temperatures, admixing a small amount of hammer-slag into the pasty mass as oxidizing blasts of air removed carbon. Heavy hammering in a semi-fluid state gave the iron tensile strength.[109] With less than 0.15 percent carbon content, wrought iron could be usefully forged, rolled or welded. It did not harden when cooled rapidly during quenching, and it resists rust.[110] Hammermen refined blooms into rods, bars or plates that smiths could work into horseshoes, gun barrels, nails, tools, agricultural implements, cutlery and tire iron.[111] But with its own insatiable appetite for charcoal and water power, the forge had to be constructed at sufficient distance from the furnace to prevent rapid depletion of these natural energy reserves.

Waterloo overlies the site of Andover Forge, built in 1761 on the bank of the Musconetcong River, seven miles southwest of Andover Furnace. In 1759, John Hackett acquired 2,164 acres in five lots, extending southwest along the summit of Allamuchy Mountain to Allamuchy Pond, Frenches Pond and Deer Park Pond. On November 22, 1759, John Hackett purchased 550 acres, lying on both sides of the Musconetcong Creek, from his father-in-law, John Reading, for £750. The Reading Tract, surveyed in 1716, extends from the head of Saxton Lake east to the Lawrence Line, including the southwestern half of the village of Waterloo. It extended southeast to about the line followed by the bed of the Morris & Essex Branch of the Delaware,

Pig iron was refined at a forge. *Charles Gillespie, editor.* A Diderot Pictorial Encyclopedia of Trades and Industry, 485 Plates Selected from "L'Encyclopédie" of Denis Diderot. *Courtesy Dover Publications.*

Lackawanna & Western Railroad (DL&W). In 1875, Seymour Smith noted the Lawrence Line, dividing East from West Jersey, "passes directly through the centre of the [Waterloo] hotel, so that while our worthy landlord is attending to his duties in East Jersey, his estimable lady can entertain her female friends in West Jersey."[112] To acquire the necessary water power, John Hackett purchased 97 acres in East Jersey on July 18, 1759, encompassing Waterloo Lake with its surface area of 68 acres.[113]

The Andover Forge Tract was completed in 1762 through purchase of 1,075 acres from the London Company, legally known as the Pennsylvania Land Company of London, an investment partnership founded in 1699 to speculate in real estate throughout Maryland, Pennsylvania and West Jersey. In 1760, Parliament placed this corporation in trusteeship, ordering the sale of its holdings and a division of the proceeds among its shareholders. On October 23, 1761, a Philadelphia auction included "2805 Acres, situate in Morris and Sussex Counties, on both sides [of the] Muskonekung River, where there is a very good Convenience for a Forge; the Land is well timbered, and a large Quantity of Meadow may be made

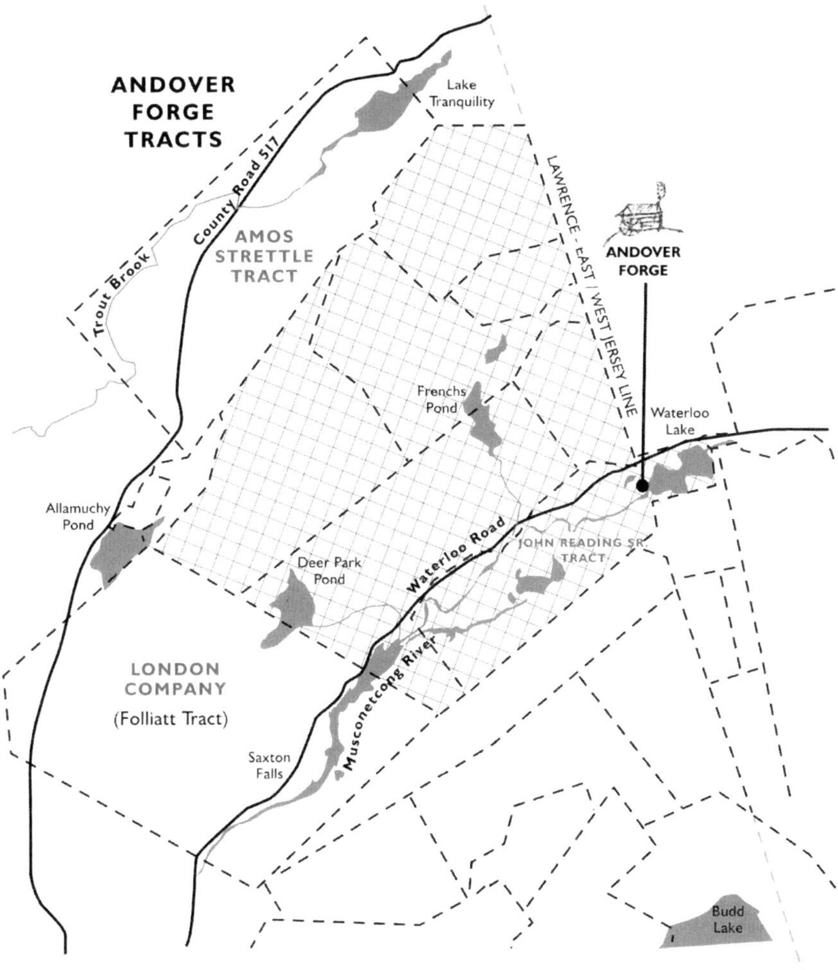

Map of Andover Forge Tracts. In 1759, John Hackett acquired five lots (2,164 acres) on Allamuchy Mountain, 550 acres on the Musconetcong Creek and 97 acres circumscribing Waterloo Lake. He bought an adjoining 1,075 acres from the London Company in 1762. *Courtesy Deborah Powell.*

of good Swamp upon it; 'tis bounded by Land of William Allen, James Alexander, and John Reading, Esquires, and others."[114] The Andover Iron Company successfully bid £1,300 to purchase 1,075 acres. On May 18, 1764, the London Company sold the remaining 1,987.75 acres to Philadelphia merchant George Bryan, who immediately resold to New York merchant George Folliott.[115]

The Andover Forge Lot grew to 3,876 acres. When all was said and done, the Andover Iron Company acquired 7,930 acres in all at a cost of £5,013. Yet the timber from this vast acreage was expended for fuel within a decade. Consequently, the owners of the new Andover Ironworks, either individually or in partnership with one another, bought extensive tracts of woodland in the surrounding countryside. In proportion to their shares in the company, the Andover partners jointly purchased a large rectangular plot of ground, containing 2,500 acres, extending from the south bank of the Musconetcong Creek in Netcong east to the ridge above the old rail yards at Port Morris.[116] George Eyre's executors sold them 890 acres in 1763, covering the hills west and south of Budd Lake.[117] After deforestation, these lands would be subdivided and sold to farmers at a handsome profit.

In 1759, the Andover Iron Company erected a log-and-earthen dam at a bend in the Musconetcong Creek, creating the necessary head and reservoir of water for its forge.[118] Despite considerable alteration of the landscape and waterways since then, sightseers could detect the tumbled remains of Andover Forge well into the nineteenth century. As late as 1859, one reporter noted, "The puddling furnaces and forge are seen on the bank of the canal at Waterloo."[119] In 1872, Edward Webb carefully determined the location of the lost industrial buildings through personal observations and interviews at the site, reporting, "The forge was situated a few yards northeast of the gristmill of Messrs. S.T. Smith and Brothers, and the walls of the old coal-house form a part of this building. While the forge was in blast there was a grist and sawmill in running order. The walls of the latter may still be seen a short distance east of the old forge site."[120] In 1875, Seymour Smith, of Waterloo, confirmed, "The grist mill of Smith Bros. is partly built on the site of the old coal house, some of the original walls forming a part of the present building. The forge and refinery were located a little to the northeast of the present mill, where remains of the same may yet be discovered. Up the stream and but a short distance from the forge, was a stone gristmill, the walls of which yet remain, a lasting monument to the credit of experienced workmanship. Farther up still, was a saw mill, but that has long since vanished."[121]

Apparently a road ran alongside the charcoal house (which the Smiths rebuilt as a gristmill), crossing the Musconetcong River on a causeway leading to the forge dam. Workmen for the iron company built a headrace or flume, using the northwest bank of the river for the inner wall and raising dirt-filled log walls, about seventy feet long and eight feet high on the riverside.[122] Head gates regulated the flow of water from the forge pond, while wooden chutes,

Dam and headrace at Waterloo propelled original mills and forge. Mound in the foreground is the foundation of the original gristmill. *Photo by Kevin Wright, 1980.*

Photograph, circa 1908, attributed to Olin Vought, shows Waterloo gristmill and saw-and-plaster mill, built on sites of the original charcoal house and forge. *Courtesy James Lee to author, 1980.*

running perpendicular to this headrace, diverted water to operate the mills and forge. Expended water discharged into the river. This headrace may partly survive at Waterloo, running along the north side of the Lock Pond—it is still used to conduct water to the grist- and sawmills.[123]

Andover Forge conformed to the English or Walloon method of forging, whereby cast iron was decarburized in a finery and then reheated at a chafery for forging.[124] The forge was a stone building, measuring about fifty by thirty feet, with four brick chimneys. Its two hammers, three smelting hearths (fineries) and one flatting hearth (chafery) produced about two hundred tons of bar iron annually.[125] The two smelting hearths worked with one hammer for fining about seventy-five pounds of pig iron at a time, for mixing the bloom and for flattening it into an ancony. About a ton of pig iron in a single charge was melted in the finery hearth under an air blast and run out into a water-cooled cast-iron trough, called a chill-mold or pit, which generally measured ten feet long, thirty inches wide and four inches deep.[126] The ancony was reheated in the chafery and flattened under a second water-driven hammer into bars.

An inventory, prepared by John Jacob Faesch and Charles Craig in 1782, mentions four bellows wheels and two hammer wheels to drive the various operations at Andover Forge.[127] Consequently, wooden flumes surrounded it on three sides.[128] A flume toward the river was fifty feet long, seven and a half feet wide and five feet high. The second flume toward the coal house, or west, was fifty feet long, five feet wide and four feet high, upheld by posts on each side spaced every two feet. The third flume toward the dam, or east, was thirty feet long, five feet wide and four feet high. The walled upper race between the two main flumes may have run along the remaining (or north) side of the building nearest the road. A bridge to the forge crossed the tailrace, carrying the road to a mill ford (later a bridge) across the Musconetcong River.[129]

The list of necessary repairs compiled in 1782 noted that a wall supported the head blocks and penstock of the Chafery Wheel.[130] The need for "a new roof over the Finery Bellows Wheel and Wall, which will take about 300 Shingles besides the lath and nails," tells us shed roofs sheltered the water wheels from snow and ice. Interior repairs included "one new poppet post, one new Anvil Block and one half of the lower part the drum Beam 50 feet long, which is two pieces, each 13 by 15 inches."[131] Each of the four hearths had a pair of wooden bellows to sustain an air blast. Immediately south of the forge, the coal house was partly of frame construction, measuring forty by seventy feet, its west gable end facing the dwelling house, which later became the Waterloo Hotel.

Fig. 78.

Finery, or run-out fire.

Cross section of finery. *From Frederick Overman,* The Manufacture of Iron, in All Its Various Branches *(1854).*

The owners commenced operations with the most sanguine expectations. On October 6, 1760, William Allen wrote John Griffiths, informing him, "Our Ironmaster here thinks three Tons, a Week, may be made at each Finery and it has been done here in one Forge, have never exceeded Seven Tons in the Chafery with a Hammer to itself. It will therefore be an Improvement if your People can draw two tons more."[132] But forging wrought iron was dangerous. Pushed gradually into a smelting hearth, one or two pigs of iron began to boil and melt.[133] A steady stream of air partially decarbonized the molten drops as they trickled into a slag bath at the bottom of the hearth.

For the twenty minutes or so that it took to melt the charge, the finer stirred and kneaded the growing iron puddle with a bar, exposing fresh surfaces to the action of the air. Puddling mixed cinder into the semi-molten iron to effect its decarburization.[134] Slag thinly coated the metal, warding off the strongly carburizing tendency of the charcoal.[135] Tongs lifted the pasty ball of iron and slag onto an anvil for hammering (*fining*) into a bloom weighing between seventy-five and ninety-five pounds.

The trip hammer worked on the principle of a lever; its hammerhead, weighing 150 to 200 pounds, was mounted to a large oak helve, see-sawing on a fulcrum. Cogs on a water-driven wiper wheel repetitiously lifted the helve, allowing hammer blows to fall. The fining hammer flattened the middle of the bloom to the dimensions of a finished bar for the length of a yard, leaving both ends unforged, so that it resembled a dumbbell. This partly worked bloom was called an *ancony*, a term derived from the French word *encrénée*.[136] Hammering altered the internal structure of the metal, compressing the crystals into an aggregate and entraining slag fibers. The ancony was reheated in the chafery with hammer scale and rich decarburizing slag before being flattened under another water-driven hammer into rods, bars or plates. Merchant bars were commonly about three inches wide, five-eighths to three-fourths of an inch thick and about twelve feet long. The entire operation for a charge of 275 to 300 pounds of pig iron took an hour and a half. Approximately 1.45 pounds of pig iron made a pound of ancony, and in flattening, 1.1 pounds of ancony made a pound of bar iron.[137] By these calculations, about 160 tons of pig iron produced 100 tons of bar iron. A normal heat, lasting about four hours, consumed 250 bushels of charcoal. Approximately 450 bushels of charcoal (representing the cutting and coaling of 1.6 acres of forest) were consumed in the manufacture of a ton of bar iron. About 330 acres of forest were coaled for the annual production of 200 tons of bar iron.

The sawmill at Andover Forge stood near the north end of the present dam at Waterloo where a concrete tumbling-vent, inscribed "1909," regulates the flow of water between the millrace and the river. A single wooden flume, fashioned with posts and planks, fed water to the sawmill and gristmill. The sawmill consisted of a heavy wooden frame forming an open shed with a wood-shingle roof, resting on wooden sills and a stone foundation. The original gristmill was a stone building, rebuilt in 1782 after suffering partial collapse when the dam burst.[138] A breast wheel, sixteen feet in diameter, provided the motive force. The mill was outfitted with a new bolting cloth in 1782. The Andover Iron Company leased the grist- and sawmills to a private operator for twenty pounds annually.[139]

Hammering the bloom. *Charles Gillespie, editor.* A Diderot Pictorial Encyclopedia of Trades and Industry, 485 Plates Selected from "L'Encyclopédie" of Denis Diderot. *Courtesy Dover Publications.*

Trip-hammer and mechanism from Hay Creek Forge, Cornwall Iron Furnace. *Photograph by Deborah Powell.*

A few houses and workshops fronted a narrow lane on the embankment above the river and mills. A wheelwright's shop overlooked the forge pond—this stone building probably became the Waterloo School House in 1842.[110] To the west, the manager occupied a two-story stone house immediately overlooking the ironworks.[111] This stone dwelling still survives: Samuel T. Smith later occupied it, adding the elaborate wooden wing in 1886. A stone dwelling house, measuring twenty-five by thirty-two feet, stood opposite the coal house—it survives as the Waterloo Hotel.[112] A large out-kitchen, measuring eighteen by thirty feet, standing adjacent to the dwelling house, was attached to the main building by addition of the taproom in 1838 and now comprises the old hearth and room in the northeast corner of the hotel.[113] Six enslaved Africans, reputedly "good Forgemen," occupied this dwelling. A smokehouse and two log stables stood behind it. Continuing west, another dwelling, known as Jack Cook's Stone House, stood on a rise of ground. It became a barn before 1782 but was returned to use as a residence in 1832 and subsequently enlarged—it survives as the Smith Homestead, opposite the canal-era General Store.[114]

Stables and a large blacksmith shop and magazine, measuring twenty-one by forty-five feet, stood near the forge and coal house.[115] North of this linear settlement were open fields, plow land and pasture. The Andover Iron Company also built a "House over the River," probably a log house located near the site of the Waterloo Ice Houses. The Andover Forge Farm covered 120 acres, including 10 acres of meadow, which rented for one hundred bars of iron per annum.[116]

In November 1762, road surveyors from Sussex and Morris Counties laid out a four-rod road from Hackettstown to Andover Forge and Gristmill.[117] Departing Hackettstown along a lane leading from Thomas Helm's Mills to Matthias Barbar's Tavern (in Hackettstown), it turned northeast through a swamp and proceeded along the river "the best way the ground will admit between the Hill and said Creek to Andover Forge Ford." It followed the "Andover Mill or Causeway and along the same to the West side thereof."[118] At the same time, a four-rod road was laid out from the road leading to Amos Pettit's Inn (Huntsville) down to the Black River at Chester. This road began near the house where Samuel Mattocks (alternately spelled Mattix or Mattox) formerly lived and then turned south along the Musconetcong Creek, following "as near to the Mill road as the Ground would admit" to the Gristmill at Andover Forge (Waterloo). This now forms a portion of Waterloo Road, between the Sussex Railroad and Waterloo. A wooden bridge, originally built at the Andover Iron Company's expense, superseded the mill ford across

The old stone portion of Samuel T. Smith's former residence at Waterloo comprises the original ironmaster's mansion. He added the frame wing in 1886. *Photo by Kevin Wright.*

Postcard view (1915) of Waterloo Hotel shows the forge master's dwelling as enlarged about 1838, incorporating the out-kitchen by the addition of a two-story wing to the right. *Author's collection.*

The corner fireplace in the forge master's dwelling (later became the Waterloo Hotel). A similar fireplace is found in the adjacent Smith Homestead. *Photo by Kevin Wright, 1980.*

Rough-hewn rafters and chimneystack in the garret of the old Waterloo Hotel, originally the forge master's dwelling. *Photo by Kevin Wright, 1980.*

Jack Cook's Stone House became a barn before 1782. The gable end reveals the original roofline and brick arch lintels of the original garret windows. *Photo by Deborah Powell.*

the river. According to Seymour Smith, writing in 1875, refined bars of iron made at Andover Forge were "conveyed down the Musconetcong valley to the Delaware River on pack horses and thence carried by water to Philadelphia."[149]

Though Benjamin Cooper was listed as manager at the Andover Ironworks in January 1764, his association with the company ended later that year when he and partner James Anderson purchased the Adventure Furnace at Hibernia.[150] Robert Taylor, employed at the Union Ironworks, succeeded him at Andover. When Richard Normand, a blacksmith who "pretends to be something of a Founder," absconded from Andover in August 1764, it was Robert Taylor who offered a ten-pound reward for his return.[151] John Hackett left as manager of the Union Ironworks and moved to Andover in April 1766.[152]

In a letter dated March 25, 1761, William Allen seemed to intimate that John Hackett suffered from poor health, writing, "Some considerable Fortunes have been made by Iron Works with us, and our present Manager [Mr. Hackett], is become an Owner with us in our New Works; and if he lives, will be a rich man."[153] Colonel Hackett was numbered among the justices and freeholders of Sussex County who appropriated £500 in May 1762 for building a courthouse and jail in Newton, serving as construction manager. He died intestate on September 20, 1766, at thirty-eight years of age, somewhat unexpectedly, for Garret Rapeljie, owner of Brooklyn Forge on the outlet of Lake Hopatcong, wrote Alexander Lord Sterling on March 30, 1767, noting he had "contracted with Colonel Hackett a few days before he died for 300 ton [of pig iron] a year, which Messrs. Turner and Allen would rather not agree to our contract."[154] In November 1766, John Hackett's widow, Elizabeth, and his administrators, Archibald Stewart, of the Union Ironworks, and Joseph Reid Jr., of Trenton, requested Hackett's creditors and debtors to settle their accounts.[155] When Elizabeth Hackett set out on a two-year trip to Europe, she and her father, John Reading, petitioned the court to appoint her brother, George Reading, the guardian for her infant son, Samuel Hackett, of Hunterdon County.[156]

By articles of agreement dated August 24, 1767, Colonel Hackett's administrators transferred his proprietary interest in the Andover Iron Company to Joseph Turner's niece's husband, Benjamin Chew, of Philadelphia, for £2,295.[157] Chew also paid the administrators £2,645 to compensate Hackett's heirs for the sum of money, "which Mr. Hackett from Time to Time as one of the Owners put into the Andover Company."[158] Chew assumed all debts due to Hackett's estate and declined interest due upon a £350 bond, which John Hackett had given him. By memorandum, Hackett's administrators agreed "that the sum of £61 2*s*, being ⅛ of £790 which Mr. Hackett has charged the Andover Company with as an advance on the Land conveyed him by Mr. Reading, whereon the Forge stands, shall be allowed to Benjamin Chew and discounted out of the Consideration Money he is to pay."[159] The widow's dower rights were extinguished. Despite these arrangements, Mr. Chew still had to defend against further claims from Hackett's estate. In 1767, he asked Joseph Reid, Hackett's administrator, and Joseph Turner to defend his title to his quarter share in the Andover Iron Company, if and when necessary.[160] In a letter dated November 8, 1767, William Allen confided, "Our late Manager…had…mismanaged our Concerns; which together with the making Iron an enumerated Commodity, has made us not pleased that we have embarked so deeply in that way."[161] On December 7, 1774, the owners of the Andover Ironworks published their intention "to apply to the General Assembly of New Jersey for an act to vest in them the legal title to sundry tracts of lands and proprietary rights, which were purchased by John Hackett in his lifetime, in trust for them and at their request, and with their money, although the deeds for the same were taken by John Hackett in his own name."[162]

Blister Steel

Andover iron was reputed to be the American ore best suited for conversion to *blister steel*, used largely for springs, saw blades, files, cutlery and edge tools, including axes, scythes, sickles, spades, shovels, hoes and plows. Made by a so-called *cementation* furnace process, from eight to thirteen tons of wrought-iron bars were "cemented" in layers, using a coarse powdery mixture of charcoal, wood ashes and salt. The bars cooked at two thousand degrees Fahrenheit for eight to eleven days in sealed firebrick chambers to exclude air.[163] External flues supplied sufficient heat for carburization at a temperature below the melting point, resulting in *blister steel*, so named because the reaction between the contained slag and carbon formed blisters on the surface of the bars. Slow cooling required another four to six days. Sometimes the steel was cut into pieces, reheated in piles and then rolled to produce *shear steel*. Repeated cutting and rolling made *double shear steel*. High-carbon tool steel resulted when blister steel was cut and melted in a covered crucible.

Steel furnaces were rare in colonial America. In his *History of New Sweden*, published in 1759, Reverend Israel Acrelius describes an early steel furnace, which Philadelphia merchant William Branson constructed on French Creek near Warrenpoint, Warwick Township, Chester County, Pennsylvania, saying it was "built with a draught-hole, and called an 'air oven.' In this, iron bars are set at the distance of an inch apart. Between them are scattered horn, coal-dust, ashes, etc. The iron bars are thus covered with blisters, and this is called 'blister-steel.' It serves as the best steel to put upon edge tools. These steel works are said to be out of operation [in 1759]."[164] Thomas

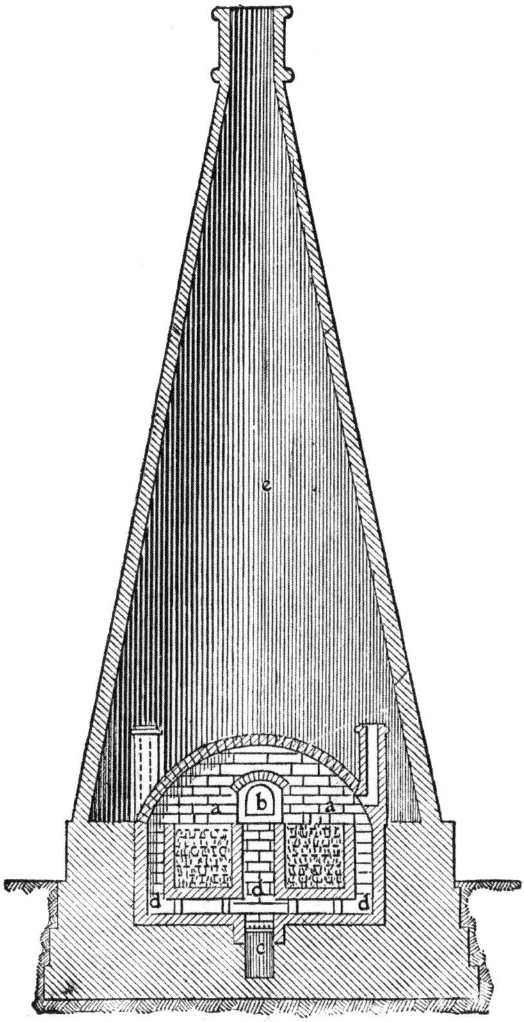

Fig. 48.* — Cementation Furnace.

Cross section of cementation furnace. *From Joseph L. Rosenholtz and Joseph F. Oesterle,* The Elements of Ferrous Metallurgy *(New York: John Wiley and Sons, Inc., 1938).*

Potts acquired Vincent Forge, near Redding Furnace, where Branson had initiated steel production in or about 1734, and revived the manufacture of blister steel there in 1765.

Isaac Harrow, an English blacksmith, established a plating and blade mill on Pettit's Run in Trenton, New Jersey, in 1734, manufacturing "axes, carpenters' and coopers' tools, tanners' and skinners' knives, spades, shovels, shears, scythes, mill and hand saws, frying pans, etc."[165] Benjamin Yard purchased "the Iron Plating Works, Smith's Shop, and all the Tools and Molds, for making Frying-pans, Dripping-pans, &c,...Also a good new Dwelling-house, Lot and Outhouses, all the Estate of Isaac Harrow, deceased" on November 1, 1745.[166] Yard apparently erected a steel furnace near his plating mill, but it was listed as unused in a 1750 inventory of such manufactures.

Benjamin Yard conveyed his steel furnace to watchmaker Owen Biddle and merchant Timothy Matlack, of Philadelphia, in 1762. They sold to Quaker merchant John Pemberton, of Philadelphia, in 1770, who offered to sell blister steel, manufactured at Trenton, and other iron products, including thin plates for making springs and mill saws and long thin bars of iron suited either to rim wagon and carriage wheels or for nail-smiths to slit in 1772.[167] Pemberton's brother-in-law, John Zane, produced steel carriage springs, scythes and saw blades on an adjacent lot in 1772. John Pemberton and Stephen Sewell offered to sell or lease "the Trenton Steel-Furnace, being a substantial stone building," in the *Pennsylvania Journal and Weekly Advertiser* on November 27, 1776. Hessians supposedly damaged the works when they occupied the city a month later. At the conclusion of the American Revolution in 1783, however, Potts & Downing were able to offer Trenton steel to English manufacturers. On August 19, 1783, a congressional committee, charged with considering "a memorial of Mr. Stacy Potts respecting a large Manufactory of Steel in which he is concerned," thoughtfully "examined many certificates from Mechanics who have made a thorough trial of Mr. Pott's Steel in edged tools and other instruments and declare that it is at least equal to the best British Steel."[168]

In a memorial published in the *Pennsylvania Gazette* on April 6, 1785, Whitehead Humphreys, a notorious conservative who publicly opposed Thomas Paine, claimed "he was the first and only person in this country who discovered and brought to perfection the art of converting bar-iron into steel, which has been considered by the best workmen and most approved judges, 'equal in quality, if not superior, to any imported into this state from Europe.'" He built a steel furnace on Seventh Street, Philadelphia, between Market and Chestnut Streets, in 1762. As early as 1766, John Hackett wrote to acknowledge a debt of £700, suggesting Andover bar-iron was being converted to steel there.[169] To encourage domestic steel production in 1770, the Provincial Association of Pennsylvania gave Humphries £100. He operated a lottery to raise additional capital in 1772.

Depression

Faced with depletion of their carefully managed timber reserves, British ironmongers craved colonial iron. Writing from London on April 14, 1763, William Allen informed Benjamin Chew, "I hope I shall have satisfied Mr. Turner that I have had thought of our great concern Iron works…I shall bring Mr. Walkers Son and some Workmen from Yorkshire and shall get Andover Iron tried in every way it can be; by all my accounts already received, which are not a few, it is the best that ever was in England. I have had several and shall have many other proposals to contract for all we can make but I shall not agree with any of them till I have conferred with the other owners."[170]

But good times for the American iron industry proved short-lived. In a creeping reversal of fortunes, the substitution of coke for charcoal steadily resurrected the English iron industry, and its output flooded American markets, depressing prices. An import duty or "enumeration" was imposed on American iron in 1764. Only about four thousand tons of American pig iron and three thousand tons of American bar iron were shipped annually to England between 1766 and 1770.[171] Capacity soon exceeded demand, and prices declined in a postwar economic depression.[172] Despite the availability of rich iron ores, American ironmakers incurred high costs for fuel and labor. As a sign of the times, one advertiser in 1767 offered to sell a tract of woodland, three miles from Andover Furnace, promising, "Timber, of which there is great Plenty, must be very valuable, being so nigh to a Furnace."[173]

In 1767, William Allen blamed the depressed iron market on British enumeration, claiming, "Most of the Forges in this and the neighboring provinces are converted into bloomeries, which will make but trifling quantities of iron, and that of but an ordinary quality."[174] He considered closing the Union Ironworks but hoped to maintain operations at Andover "for a year or two to see if times will mend, which is not a little mortifying as we have an inexhaustible bank of ore and the iron not exceeded in quality by any in America, and I might venture to say, not even in Europe." In September 1768, Allen complained those ironworks remaining in operation "sink money to the owner every year, and they will shortly be reduced to only as many as can manufacture for ourselves and that will be the case with almost every article of our consumption."[175] On November 7, 1769, Allen again blamed imperial trade policies for the economic downturn, writing, "Indeed I can assure you that property is lowered near half its value in most of, if not all North America; particularly Iron Works, in which Concern many people by the Encouragement from England, in taking off the Duties, had of late embarked, have their works in a manner knocked up. The enumerating of that Commodity has put most of our Works to a stand, and Mr. Turner & I intend to let ours, which is one of the most considerable in America, lie still till better times."[176]

Archibald Stewart became superintendent of the Andover Ironworks and moved to Andover Furnace in 1769.[177] Robert Taylor succeeded him as manager of the Union Ironworks. On August 4, 1770, Stewart offered a three-pound reward for the capture and return of John Collins, a runaway English servant, twenty years of age. He was described as "5 feet 5 or 6 inches high, round shouldered, full faced, a little freckled, has short straight brown hair, slow in speech, a little dull in hearing, stoops and rocks much when he walks; had on, when he went away, a drab colored coarse cloth jacket, with sleeves, one blue double breasted under ditto, without sleeves or lining, metal buttons on both, an ozenbrigs shirt, and pair of trousers, an old felt hat, shoes half worn, with buckles not fellows."[178] But this was not the worst of Stewart's problems. Squeezed by high operating costs and low returns, the owners offered to lease the Andover Furnace and Forge in October 1770. They included "six Negro Slaves to hire out or sell, who are good Forgemen, and understand the making and drawing of Iron well."[179] In 1771, Archibald Stewart acquired 150 acres on the northwest branch of the Pequest Creek, where he established Springdale Mills in what is now Andover Township.[180]

On May 15, 1772, Lynford Lardner, a junior partner, leased his quarter share to William Allen, Joseph Turner, Benjamin Chew and Archibald

Stewart.[181] By his last will and testament, dated September 13, 1774, Lardner named his wife, Catherine, and son John as executors. John Lardner, of Philadelphia, advertised the sale of "one equal and undivided fourth part of the Andover Ironworks" in the *Pennsylvania Gazette* in 1775. Again, in May 1777, Lewis Weiss, of Second Street, Philadelphia, offered to sell Lardner's quarter-share interest, including nearly twelve thousand acres of land and an inexhaustible body of ore.[182] Operations persisted. James and Alexander Stewart advertised the sale of Andover pig and bar iron on Cruger's Wharf, fronting the East River, in the *New-York Gazette and the Weekly Mercury* on June 27, 1774.[183] Sporadic accounts of the Andover Iron Company, partly preserved in Joseph Turner's ledgers, record business activity in March 1775, May 1776 and February 1778.[184]

Revolutionary War

William Allen offered cannon shot manufactured at his ironworks to the Philadelphia Council of Safety in October 1775, but his enthusiasm soon dampened. Anticipating General Howe's imminent capture of Philadelphia in September 1777, Judge Allen suspended operations at Andover. By articles of agreement, dated September 20, 1777, he and Joseph Turner agreed to lease "Eight Negro men Slaves, Viz: Tom Pipes, Harry, Mingo, Caesar, Peter, Burrouss, Jack Martin and George Jack" to John Patton, Esq., of Bucks County, Pennsylvania, for one year beginning October 1.[185] They were to be employed as laborers "under his direction in any Reasonable Service about said Patton's Ironworks." Patton agreed "to provide for said Negroes Sufficient Meat drink Washing and Lodging, together with one upper and one under warm Woolen Jacket, four new Shirts, two Pairs Trousers, one Pair Breeches, two pair Woolen Stockings, and three pair Shoes for each and every one of the above Negroes."[186] He was not to let any of them suffer for want of a doctor in case of illness. At the conclusion of their term of service, Patton was to return them to the Union Ironworks, paying thirty-five pounds each for Tom Pipes, Harry, Caesar and Peter; thirty pounds for Burrouss; twenty pounds for George Jack; and ten pounds for Jack Martin.

General George Washington appointed Lieutenant Colonel Benjamin Flower, commissary general of military stores, to organize and command an American regiment of artillery artificers whose duties were to gather and distribute arms and provisions, cast cannon, bore guns and manufacture ammunition.[187] Already burdened with administering his new department,

Weighing pig iron at furnace. *Charles Gillespie, editor.* A Diderot Pictorial Encyclopedia of Trades and Industry, 485 Plates Selected from "L'Encyclopédie" of Denis Diderot. *Courtesy Dover Publications.*

Flower was overstrained by the sudden task of evacuating military stores from Philadelphia as British troops approached the city. Soon thereafter, he "was taken sick at the Valley Forge with the Nervous Fever, [and] was confined [to his bed] near two months." Outmaneuvering his opponents, General Sir William Howe occupied Philadelphia on September 26, 1777.

On January 15, 1778, the Board of War and Ordnance sought congressional authorization for Colonel Flower to contract with Whitehead Humphreys "for making a Quantity of Steel for the Supply of the Continental Artificers and Works with that necessary Article, and as the Iron made at the Andover Works *only* will with certainty answer the Purpose of making Steel, Col. Flower be directed to apply to the Government of New Jersey to put a proper Person in Possession of these Works (the same belonging to Persons who adhere to the Enemies of these States) upon such terms as the Government of New Jersey shall think proper."[188] Congress concurred and forwarded the board's request to Governor Livingston, the letter "setting forth the peculiarity of the demand for these works, being the only proper means for procuring iron for steel, an article without which the [military] service must irreparably suffer."[189]

Though warned against resuming his duties prematurely and risking a relapse, Commissary General Benjamin Flower departed camp on February 22, 1778, for Allentown, Pennsylvania, where he supervised the removal of military stores to Lebanon, and then traveled to New Jersey to deal with the matter of the Andover Ironworks. Recognizing "that the Andover iron is *better suited to this business* than any other in America," Governor Livingston urged the New Jersey legislature to act, saying, "I doubt not you will readily comply with the expectations of Congress in this respect."[190] Despite this encouragement, the governor's council instead wrote to Lieutenant Colonel Flower on March 18, 1778, suggesting Colonel John Patton, of the Ninth Pennsylvania Regiment, should try to lease the Andover Ironworks from its principal owners, William Allen and Joseph Turner, who were then in Philadelphia behind enemy lines.[191] Colonel Patton and Whitehead Humphreys, the Philadelphia steelmaker, instead met with Loyalist Benjamin Chew, former chief justice of Pennsylvania, who was then residing under parole at Solitude House in High Bridge, Hunterdon County, at the Union Ironworks. Chew "refused to have anything to do with it, as he was only a part owner [of the Andover Ironworks], but advised me to send in a flag [of truce] to Philadelphia, in order to treat with Messrs. Allen and Turner on the terms of the lease…which advice of Mr. Chew I considered as an insult, as he knew such a step was impracticable."[192]

Unable to come to a satisfactory arrangement with Colonel Patton, the Board of War instructed Colonel Flower on May 25, 1778, to instead "procure the possession of the Works for Col. Thomas Maybury, with whom the terms of the contract are settled, on condition of his getting the possession agreeably to the resolution of Congress."[193] Accordingly, Colonel Flower petitioned Governor Livingston on May 28, 1778, to "procure the possession of the Works, to wit: the furnace and forges for Col. Thomas Maybury, with whom I have made a contract, for the iron to be made at the said Works, to be converted into steel."[194] On May 25, 1778, Richard Peters, secretary to the Board of War, also wrote to Livingston, emphasizing, "As we find it absolutely necessary to put these works in blast, the Board beg

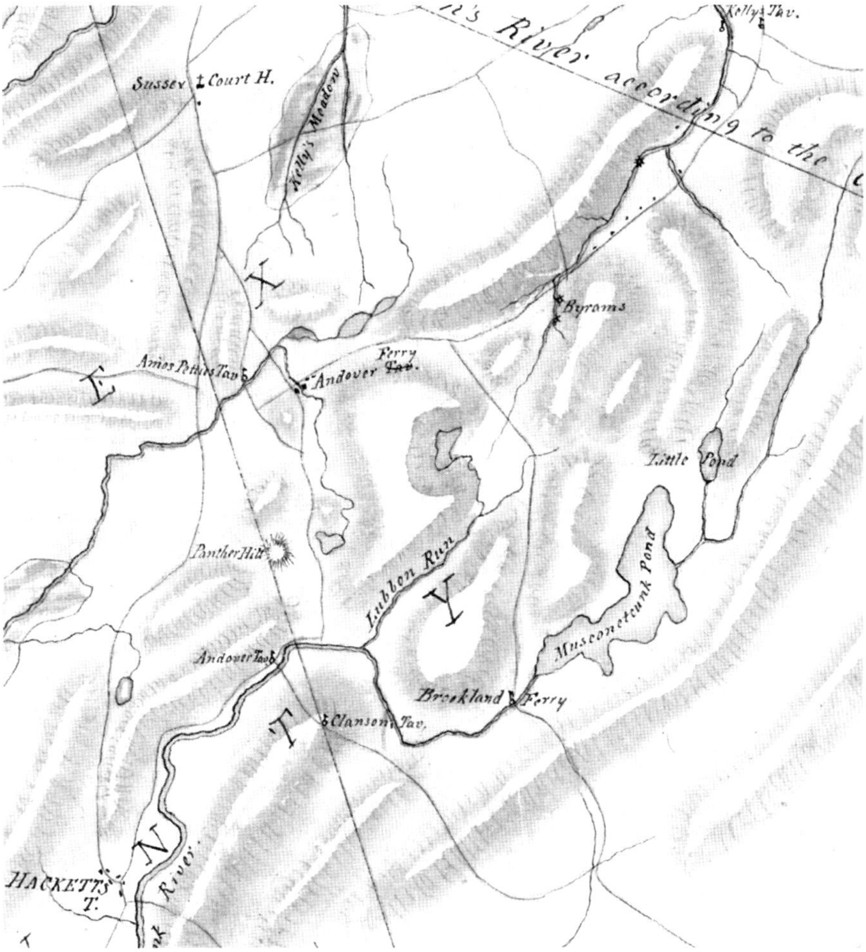

Detail from "Three maps of northern New Jersey, with references to the boundary between New York and New Jersey," [1769?], Library of Congress. Andover Furnace is misrepresented as "Andover Ferry" and Andover Forge is identified as "Andover Tav[ern]." The town of Newton is identified as "Sussex Court H[ouse]."

the favor of your Excellency to assist Col. Flower or Col. Maybury in the business."[195] On June 1, 1778, the governor asked the General Assembly to intervene in order to secure "so essential an Article as that of Steel in the Army of the United States."[196] "An Act to empower certain Commissioners therein named to take Possession of and lease out the Andover Iron-Works, in the County of Sussex," was signed into law on June 20, 1778, establishing a three-man commission to control the works for three years.[197]

Colonel Thomas Maybury agreed with Colonel Flower to carry on the Andover Ironworks on May 28, 1778, for the sum of $13,333, but he apparently awaited the appointment of the New Jersey commissioners, because his accounts show he first manufactured three hundred tons of pig iron at £73.6.8 per ton at Andover in July 1778.[198] On August 19, 1778, Joseph Nourse, paymaster to the Board of War and Ordnance, advanced the sum of £2,253 ($6,000) to Colonel Thomas Maybury, of New Jersey, under his agreement with Colonel Flower on a contract for iron, dated at Lebanon, May 28, 1778.[199] Maybury produced only one ton of pig iron in October 1778 at a cost of £80 and another thirty-three tons in November 1778 at £83.6 per ton. On October 15, 1778, he received twenty-seven barrels of flour, valued at £113.1.3, in partial payment on his contract. On November 3, 1778, Maybury was paid £2,750 in cash. Colonel Maybury delivered two tons of Andover pig metal at Ancocus Old Long Bridge Landing at £500 per ton on February 27, 1779.[200]

In an advertisement posted in the *Pennsylvania Packet* on July 9, 1778, steelmaker Whitehead Humphreys offered to reward anyone who could provide information on the whereabouts of "four tons of Blistered steel, with the apparatus and all the utensils belonging to the subscriber's steel furnace," which Joseph Fox, a Loyalist, seized and "sold to some persons in town," shortly after the British occupied Philadelphia. Disaffected citizens at the Andover Ironworks also provided aid and comfort to the enemy. Writing to George Washington on January 9, 1779, Governor Livingston reported, "Thirty odd of the [General Burgoyne's] Convention troops have deserted in a Body at Andover." Near twenty of them were retaken and imprisoned at Morristown, though five again escaped, only to be recaptured at Newark. Two of the escapees implicated two inhabitants of Sussex County, "who induced them to desert at Andover, and aided, supported and piloted them part of the way in their intended march to New York."[201] Because of these and perhaps other difficulties, Colonel Thomas Maybury made a little over six tons of pig iron at Andover in January 1779 at a cost of £166.13.4 per ton; twenty tons in February 1779, which cost between £173 and £175 per ton; and another three tons in April 1779 at a cost of £300 per ton. To sustain operations, he received 113 barrels of flour on January 3, 1779; 100 barrels of flour and 50 head of beef cattle on February 5; and 145 bushels of wheat on April 1. The Board of War informed Congress on April 1, 1779, "that with every assistance they could afford the Andover works, which are now carrying on under contract with the United States, the business goes on very tediously and heavily, by reason of the great difficulty in procuring hands, and other impediments, which would be likely to happen in any other works under the present circumstances of the Country."[202]

In August 1779, Colonel Maybury produced over seventy-four tons of pig iron, but the cost now ranged from £300 to £466 per ton. In partial payment, he received 238 bushels of wheat on August 1, 1779, and forty-seven barrels of beef on August 23. In November 1779, Maybury made four tons of pig iron, receiving forty-seven barrels of flour on November 13, 1779, and another eight barrels two days later. He received cash payments of £15,000 on August 23, 1779, and £5,000 on November 9. On January 17, 1780, Colonel Maybury shipped twenty-five tons of pig iron, produced at a cost of £500 per ton. He received another £5,000 in cash, leaving an outstanding balance of £43,973.1.8. On March 12, 1780, Maybury hauled two tons of pig metal to the Trenton forge and one ton and fifteen pounds of bar iron to Cooper's Ferry at a cost of £1,875 per ton.[203] Despite these deliveries of pig iron, Major Samuel Hodgdon prepared a report from the Board of War to Congress on December 2, 1780, which included the purchase of "500 Weight of German Steel @ £18.15" per hundredweight for the use of the Regiment of the Artillery Artificers.[204]

Thomas Maybury and James Morgan petitioned Congress on March 17, 1780, for payment of £61,473.1.8, the balance they thought due on their contract with the Board of War.[205] The president of Congress referred their memorial to the Board of Treasury.[206] On March 21, 1780, the Treasury Board reported, "They have considered the petition of Thomas Mayberry and James Morgan, Iron Masters, and cannot discover that any thing therein comes within the description of Balances of accounts liquidated, of partial payments of such accounts, or of advances for public service; and therefore humbly conceive it is not regularly within their Department to make any report thereon, and that the matters therein contained are proper for the consideration of other Departments who are conversant with the Ordnance supplies."[207] Richard Peters, writing on behalf of the Board of War and Ordnance, now informed all concerned, "At a time when the public were in great distress for the Want of Steel, the Andover Iron Works (which produced the only Iron in America which with the most certainty can be converted into that Article) were delivered to Col. Maybury by the Legislature of New Jersey & a Contract was made with him for a Quantity of Pig Metal."[208] He further noted, "The public complied with their part of the Agreement in due Time, the Case was otherwise with Mr. Maybury who delayed the Performance of his Part of the Contract for an unreasonable Length of Time. Finding little Prospect of Advantage to the public from a Continuance of the Contract with Mr. Maybury, the Board thought it Best to put an End to the Business & directed the Commissary General of Military Stores to Settle the Account with that Gentleman." However, "on finding the Balance so large & knowing the public Embarrassments & judging that all the Pig Metal would not be indispensably wanted, we agreed to deliver fifty Tons to Mr. Maybury at the Price the public gave him for it, which being £500 per Ton, would have reduced the Balance considerably, but Regulations of Prices being then apprehended or for some other Reasons, Mr. Morgan, who acted in behalf of Mr. Maybury, refused to take above half the Quantity we offered him. We are still waiting to comply with our Offer if Mr. M. will accept it & as to the Balance we are of the opinion it should be paid as Mr. Maybury is a Sufferer as to Hurt by Depreciation. Whether it comports with the public Circumstances to pay this Balance is best known to the Treasury Board. We beg leave therefore, after having stated our Knowledge of the Affair, to respond That the Memorial of Messrs. Maybury & Morgan, Iron Masters, asking Payment of the Balance of their Agreement, together with a Letter from the Board of War & Ordnance on that Subject be referred to the Treasury Board." Messrs. Maybury and Morgan responded with a letter on March 22, "asking payment of the balance of their account," enclosing the letter from the Board of War, which was referred again to the Treasury.[209] With this documentation in hand, Congress ordered Thomas Smith, commissioner of the Continental Loan Office for the State of Pennsylvania, to pay $117,261.69 in loan office certificates, charged against the account of the commissary general of military stores, to Colonel Thomas Maybury.[210] On March 28, 1780, Colonel Benjamin Flower certified that he settled accounts with Colonel Thomas Maybury, paying a balance of £5,000 in Pennsylvania currency.[211] In an estimate of the "Shells & Shot cast" by requisition of General Henry Knox, dated March 1, 1780, Thomas Maybury reportedly produced twenty tons at Taunton Furnace in Medford, New Jersey.[212]

According to Commissary Samuel Hodgdon's accounts, Colonel Thomas Potts, ironmaster, received five shipments of Andover pig iron between April 11 and May 23, 1780, totaling about six tons, for conversion in his blister steel furnace, located on French Creek in Chester County, Pennsylvania. Between September 21 and October 31, 1780, Colonel Potts received another twelve shipments, totaling about thirty-seven tons. Additional records show Colonel Potts received one ton of "Andover pig metal," delivered at Philadelphia, on November 2, 1780.[213]

Judge William Allen died on September 6, 1780. Acting on behalf of "the owners in Philadelphia," Archibald Stewart, of Hackettstown, offered to lease the Andover Furnace and Forge in the *New Jersey Gazette* on November 8, 1780, indicating that "a large quantity of wood ready cut may be had convenient to the works" and offering to sell "some Negroes belonging to said works."[214] Upon Colonel Benjamin Flower's death on May 1, 1781, Samuel Hodgdon succeeded him as commissary general of military stores and assistant quartermaster.

Somewhere in this interval of time, a natural disaster intervened to halt the operation of Andover Forge. Assessing repairs needed there in June 1782, John Jacob Faesch and Charles Craig reported the dam had washed out, opening a breach eighty-nine feet long.[215] Floodwaters swept away the head gates, undermining a wall about seventy feet long and eight feet high, which formed the lower side of the headrace. The rush of water damaged the flume to the grist- and sawmills. The stone wall of the gristmill toward the dam collapsed into the tailrace, irreparably injuring its water wheel. To show the height and fury of the deluge, the gable end of Colonel Potts's stone house "as far as the upper floor" was down, and the dwelling needed a "whole under floor and half an upper floor." The bridge across the Musconetcong River "was carried away when the dam broke and has been built at the Company's Expense on account of the road not being laid in Sussex, a good deal of Timber which is down the river might be brought back, all the planks are wanted." Three flumes serving the forge were destroyed, and the six water wheels needed to "be sealed and repaired." A bridge over the tailrace to the forge was swept away. The assessors could not even inspect the condition of two pairs of tub bellows, but "as they suffered a great deal in the Water when the dam broke, the repairing of them will therefore cost a good deal."

Whether or not this setback was the cause, Colonel Thomas Potts was slow to deliver. Samuel Hodgdon wrote him from Philadelphia on March 27, 1782, complaining, "It is a matter of surprise to me that so small a part steel for which you have long since been furnished with Metal, has been received. We are now in want, must beg the favor of you to furnish One Ton (at least) immediately, and at the same time forward an account of the quantity of Steel delivered at my store, or to my Order, that I may know what more I have coming agreeable to Contract."[216] Colonel Potts eventually produced about three tons of blistered steel for the government, which he delivered in various amounts on May 22, September 20, October 7 and December 20, 1780; and on April 11, 1781; May 6, 1782; and May 20, 1785.[217]

Time-consuming and expensive repairs to Andover Forge forced Hodgdon to contract with other forge owners along the Musconetcong River to refine Andover pig metal into bar iron. He also turned in frustration to other steel manufacturers, informing Colonel Jacob Weiss, assistant deputy quartermaster-general at Easton, Pennsylvania, on July 12, 1781, that he

> *entered into Contract with Messrs. Potts & Downing of Trenton upon certain Conditions (therein expressed), to covert the whole of the Pig Metal at Easton, & its vicinity, into good blistered Steel, agreeable to the intention of the Public, when the Pigs were procured. I have draw'd an Order on you for the delivery of the whole, as circumstances shall make convenient, the Pigs by contract are to be transported to the most Contiguous Forges for drawing, and the Barrs when drawn are to be sent off to the landing near the furnaces at Trenton at the expense of the* [United] *States, which I am in hopes, the Conveyance being mostly by water, will not be attended with any great expense. The Steel when made is to be sent to Philadelphia at the expense of the Owners of the Works, or if taken from the works, the expense of transportation allowed, the exchange as stipulated by Contract, under the above Cited Circumstances, is Seven Tons of Pig Metal for one Ton of warranted Steel, which appears to be advantageous to the United States.*[218]

He also ordered that pig metal and bar iron belonging to the government at Colonel Mark Thompson's Changewater Forge (Washington Township, Warren County, New Jersey) be forwarded immediately to replenish stockpiles.

But things did not go any better in Trenton. On March 24, 1782, Hodgdon pleaded with Colonel Weiss to "be so obliging immediately upon the receipt of this Letter, as to send me an account of Barr Iron & Pig metal, delivered by you or your Order to Messrs. Potts & Downing in consequence of my Letter some time since, the Contract made with them being Shamefully violated on their part, in justice to the Public. I must suspend any further advance of Metal or Barr Iron, until they shall have accounted for what they have already received agreeable to the Spirit & Letter of the Contract, whether they received the Barr Iron from you or Colonel Thompson."[219] In utter frustration, Hodgdon was determined to settle accounts with those who were contracted to convert Andover iron to bar iron and for "procuring a Quantity of Steel which is now Greatly wanted."[220] In near fury, he complained, "Not withstanding the ample & Expensive provision made by the public to procure metal suited to their purposes & the generous terms held out in the several Contracts made to carry their views into execution, such has been the Construct of those Concerned, that the Country might have been lost for want of that article which they have been at so much cost & pains to provide, I mean Steel—."[221]

Joseph Turner and the other owners leased the Andover Ironworks to Archibald Stewart for three years, starting November 1, 1782.[222] They were "to be concerned with him in the Profits and Loss in carrying on the works during the lease." A teamster identified as L. Willson was paid for carting a ton of Andover iron to Trenton in 1783, indicating a revival of operations.[223] In 1784 and 1785, company ledgers indicate at least six tons of Andover iron were carted to Trenton, eight tons to Philadelphia and four tons to Elizabeth.

Joseph Turner died a bachelor on July 25, 1783, aged eighty-two years, devising his estate to his nieces, Elizabeth and Margaret. Elizabeth became the second wife of Chief Justice Benjamin Chew. Margaret first married sea captain James Oswald, and after his death, she wed Frederick Smythe, chief justice of New Jersey. In 1784, Mary Allen, widow of John Allen, petitioned the New Jersey legislature for authorization to sell the interest of her minor children in the Andover Ironworks, stating that her children inherited a $^{5}/_{16}$ interest devised by William Allen to his grandsons.

The American Revolution placed new demands on New Jersey iron furnaces, especially for munitions production. With reduced competition from imports, the domestic manufacture of cast-iron ware increased considerably.[224] Yet productive capacity declined. Many merchant capitalists, residents of Philadelphia and New York, who established the largest ironworks, were suspected (if not outrightly accused) of loyalty to the Crown. Despite exemptions from military service, a scarcity of skilled workers aggravated already high labor costs. In some instances (as at Andover), the military coercively "leased" ironworks but made some compensation to their owners. Some ironworks were discontinued as forced production consumed the forests surrounding them.[225] As Johann David Schopf, a German observer, noted in 1777, "The mining and metallurgical industry in New Jersey, as everywhere in America, cannot be enduring in its present condition because no care is taken, as is done in most districts in Europe, to maintain the forests, and many works must stop without uninterrupted supplies of coal and timber, as is here and there already the case."[226] According to Schopf, the ironmasters "cut over [large tracts of woodlands] without order." The Union Furnace in Hunterdon County, having consumed forty-two thousand acres of timberland in fifteen years of operation, was abandoned. Even after this cleared land was settled in farms, its occupants suffered for want of wood.

Peace Returns

In the absence of protective tariffs, peace brought a resurgence of cheap foreign imports. High costs and low prices made it difficult for American manufacturers to compete in their own coastal cities. Carting iron products from Highlands' furnaces to tidewater markets in New York or Philadelphia cost approximately £1.15*s* per long ton.[227] Of an estimated 3,500 long tons of pig iron produced in New Jersey in 1783, only 800 long tons were sold as pig iron or cast-iron wares.[228] The remaining 2,700 long tons were converted to bar iron at local forges. In 1783, pig iron cost £6 per long ton to make and sold for £8 to £9 per long ton at Philadelphia. Bar iron cost £22.2*s* per long ton to make but sold in the same market for £27 to £30 per long ton.[229]

In 1785, the English economy slumped, and the price of bar iron at Philadelphia dropped from a temporary high of $112 per ton in 1784 to $68 per ton in 1786. Acknowledging that Whitehead Humphreys supplied not only Pennsylvania but also most of the United States with steel during the American Revolution, the Pennsylvania legislature approved a £300 loan in April 1786 to allow him "to carry on the manufacturing of steel within this state, he giving them satisfactory security for the repayment thereof within five years, without interest."[230] By 1790, bar iron crept back to about $80 per ton, signaling an improved market.[231] Despite such fluctuations, British ironmongers expanded capacity from thirty thousand tons in 1775 to eighty thousand tons in 1790. Their success was largely founded upon technological innovation, especially the conversion to coke and introduction of the Watt steam engine to drive the air blast. In contrast, many Highlands iron plantations, starved for fuel, were sold off as farmland. In 1789, Assistant Treasurer Tench Coxe counted eight furnaces and seventy-nine forges operating in New Jersey, only two-thirds of the number active before the Revolution.[232] Local demand rather than foreign markets largely sustained New Jersey's iron industry as furnaces sold pig iron to independent forges and small foundries in their neighborhood. These in turn supplied country blacksmiths and mechanics with wrought iron. Bar iron even substituted for more precious metals as a currency in country trade.

The Andover Ironworks barely clung to life in this inhospitable economic climate. On February 27, 1784, Archibald Stewart complained to Philadelphia merchants Peter Kuhn and Gustavus Risberg, "The Winter has been so very severe that we have done very little at the Forge, the wheel being for the most part of the time froze fast."[233] He was sorry the stock of iron at Trenton could not be transported across the frozen Delaware River. He wrote again in June, this time protesting that they had "been much hindered from working in the Forge by the breaking of Hammers."[234] He and his employees stood "in want of Some Pork" and wished the firm "to purchase ten, fifteen or twenty bars [of iron]" in order to enable them to get provisions.

In December 1784, Benjamin Chew Jr. informed his father that Charles and Archibald Stewart were

> *desirous of taking the* [Andover] *Works and opening an extensive Business from which they draw the most flattering Expectations. The Iron is in high Reputation and much in demand, both daily increasing under the present Tenant, in so much that when other Iron has been sold at £35 per Ton, Andover sells at £50. The Lease of the Man who now rents it expires in March. Charles Stuart* [sic] *is about leaving his Military Employment and will either rent the Works for himself and a company* [or] *hire or engage with Archibald Stewart in a Concern with you in any Share you Choose to retain—the latter first proposed the Plan and seems well convinced much Money may be made if carried thro' under proper management—he would overlook the whole & put the Wheels in Motion under the direction of a Man brought up under him adopting at the same time many different Regulations from the former mode of conducting Business.*[235]

They felt that £1,500 to £2,000 in capital stock would be required to carry on the scheme "in the extensive manner they propose."

In May 1785, the surveyors of the highways for the townships of Independence, Hardwick and Newton agreed to lay out a four-rod road on

Postcard shows Andover Furnace, which Joseph Northrup rebuilt as a gristmill in 1816. Fire gutted it in 1911. The stone building to the right may be the original blacksmith shop. *Author's collection.*

the old Indian trail across Allamuchy Mountain, departing from the highway (Route 517) between Hackettstown and Newton, near Moses Wilson's house and about 660 feet from Alamuche Pond Brook, running across Allamuchy Mountain to Andover Forge (Waterloo), almost where Interstate 80 now crosses. On May 23, 1785, surveyors continued this road across the Andover Forge Tract, running beside the dwelling house (later the Waterloo Hotel) and near the coal house (later the Waterloo Gristmill) to the river, where a bridge crossed the Musconetcong River into Morris County.[236]

Andover Furnace was put in blast on June 25, 1785, and produced 220 tons of pig iron before it was blown out on October 11, 1785.[237] After putting in a new hearth, the furnace resumed operation on November 13, 1785, producing 346 tons before its fires were extinguished on May 8, 1786. In June 1785, two New York merchants, Samuel Ogden and Thomas Bridgen Attwood, offered to sell, wholesale or retail, bar iron manufactured at Andover, Mount Hope and Boonton.[238] In January 1786, the Stewarts sold 100 tons of pig metal to John Stotesbury, who operated the Greenwich Forge along the Musconetcong River at Warren Glen.[239] Thus they carried on the ironworks beyond the term of their original lease.

Benjamin Chew traveled to Andover in the spring of 1786 to review Archibald Stewart's ledgers but "in the weak State he was then in would not admit of my troubling him further than barely to mention the matter to him and he was kind enough to promise me it should be done as soon as his Health & a little Leisure would enable him to do it."[210] Chew received no further word from Stewart over the summer. On August 20, 1786, Benjamin Chew asked Robert Taylor "to prevail with my good friend Archibald Stewart to fix a time to bring those Books down at least as far as Union [Ironworks] where I would meet him."[211] Chew apparently did not suspect chicanery but, on the contrary, continued to believe Stewart "to be a good Man and as such I have the most sincere Friendship for him & I am persuaded that the busy Scenes he has been involved in at Andover must have engaged his whole Attention and when he has time."[212] In February 1787, Chew learned from a visitor "of another Severe Attack of Illness on poor Archibald Stewart" who was "deprived of [his] faculties of reason."[213] As executor of Joseph Turner's estate, he "went to see Archibald Stewart at Sussex Courthouse [now Newton] & found him in an estate of almost total insanity—but I brought him down with me to Andover Forge, where Charles Stewart then had the management of the Works." He despaired of settling Turner's estate, saying, "The greatest Obstruction to which is the unsettled Situation both of his own and Allen and Turner's Jersey Concerns—These Matters can never be adjusted by the Books Mr. Turner kept here, but if they are ever to be ascertained it must be done by the Books and Accounts kept at Union and Andover."[214] According to account books, at least thirty-two tons of pig metal were carted from Andover Furnace to New York in 1787 and another thirty-three tons of bar iron from the forge.[215]

In January 1788, Major William Helms, Hackettstown merchant, presented his account against the Andover Ironworks, "authenticated by a Settlement and signature at the End of them in the name of Archibald Stewart & Company."[216] He demanded satisfaction from Benjamin Chew, who refused to pay charges he considered exorbitant. Instead, Chew asked his agent to find out whether "William Helms has signed for Iron taken by himself without written order" between June 1785 and May 1786. He also wanted to know from the Blast Book how long the furnace had been out of blast in the fall of 1785, whether he had ever "conversed about the Lease or if he knew of it" and whether Helms sold iron on credit. According to Chew, Helms sold iron on several occasions on his own account and "yet charges Andover Forge with the loss in weight that happens thro' his own or his Servants' Negligence." Records showed Helms received shipments of bar iron and nail rods from Andover Forge in 1786 and 1787.[217]

Troubles continued to mount. In January 1788, Benjamin Chew sent Archibald Stewart a copy of a letter he received from Daniel Smith, of Burlington, "respecting some wood he says was cut on his Land near Andover Forge, with Draught of that Part of the Land on which the Wood or Timber stood as I am totally a Stranger to the matter & you best know whether he has or has not sustained an injury & to what amount."[248] Chew recalled that Stewart had "informed me that you believed the Wood cutters had by mistake gone a little over the Line, but it being very soon discovered they were stopped after cutting a small Quantity on his Land."[249]

On March 8, 1788, the sheriff of Philadelphia delivered summons to Archibald Stewart, Charles Stewart, Benjamin Chew Sr. and Robert Taylor to appear and answer the complaint of William Helms in a plea of trespass. They reached agreement on March 22, 1788, "concerning suits to be grounded on a partnership supposed to have Subsisted" among them.[250] Merchants Peter Kuhn, Gustavus Risberg and Daniel Cahill Jr. were subpoenaed to testify in Trenton on November 9, 1789, regarding business dealings between Archibald Stewart and Messrs. Kuhn and Risberg, of Philadelphia, from July 1783 to September 1784.[251] On November 9, 1790, the Supreme Court required Archibald Stewart to bring with him the "Day Books, Waste Books, Journal Books, Blast Books, Ledger Book" and especially any accounts for the period of his lease from Joseph Turner, beginning November 23, 1782.[252] The records apparently showed careless management and deliberate abuse by different agents of credit with Kuhn and Risberg.[253]

John Armstrong leased Andover Forge with its sawmill and gristmill for three years in 1788 for an annual rent of eight tons of bar iron, to be taken out of the first bars that were made at the rate of every fourth ton until the rent was discharged.[254] Armstrong agreed to keep the works in full repair and to surrender the premises, either when the owners accepted a rental offer on the whole ironworks or in the eventuality a division of the property was made among the owners. Frederick Smythe, one of Turner's devisees, approved the lease on April 1, 1788. Confusion prevailed, however, for Benjamin Chew learned in March 1789 that Mr. Armstrong had been at work for some time at Andover Forge without his knowledge.[255]

Abraham Baily wrote to John Armstrong on September 28, 1789, reminding him that he had promised "to pay the Rent agreed upon for the Andover Forge at the Rate of every fourth Ton of Bars of those first made until the Rent was discharged."[256] He informed Armstrong that the owners had recently agreed to rent the forge and furnace to Messrs. Baily, Hill & Evans, who wished to take possession immediately. On June 30, 1789, Benjamin Chew ordered Armstrong to deliver two and a half tons of bar iron to Benjamin Chew Jr., the same to be allowed out of the rent due him.[257] Archibald Stewart went to Andover Forge in January 1790, seeking Armstrong, who was not there.[258] He reported to Robert Taylor, "The principal thing of value there is the [char] Coal, supposed better than 300 loads."[259]

On October 29, 1789, Benjamin Chew and his wife, Elizabeth, of Philadelphia, leased their interest in the Andover Ironworks to Dr. Abraham Baily, Humphrey Hill and Cadwalader Evans of the Counties of Delaware and Chester, Pennsylvania, for five years, commencing June 25, 1790.[260] The lessees were to pay £84 7*s* 6*d* for the first year and £168 15*s* for the remaining years. They were also to pay any taxes and assessments on the 8,122 acres, receiving a commensurate abatement in their rent. Baily, Hill and Evans further agreed to "carefully husband and preserve the wood and timber on the Premises," making "their best Endeavors to purchase and procure Cord wood from the Lands of the Neighborhood," for which Benjamin Chew agreed to deduct six pence for every cord purchased from his share of the rent.[261] The leaseholders also promised to maintain "such Lands as may be in Cultivation in a Farm-like Manner." While forbidden from removing any ore or cinders found on the premises, the lessees were allowed to pick, sort and stamp the cinder heap at the furnace for a period of two months prior to June 25, 1790, and to use the stamped ore at Andover Forge so "that the said Forge may be put in Motion and some Profit made thereof by the said Lessees…and before the Furnace before mentioned can be put in Blast to supply the said Forge with Pig Metal."[262] The new operators were to pay 20 shillings to Benjamin Chew for every ton of bar iron wrought at Andover Forge.

Since the new lessees feared "some present Defect in the walls of said Furnace," it was agreed that "if upon experiment made in first Heating the Furnace or within 6 weeks from beginning to heat the same, it should clearly and evidently appear without any default in overcharging the said Furnace or other Mismanagement of [the] Lessees or their agents that the Stack is so injured or in such ruinous State as not to be fit to be put into Blast," then Benjamin Chew would bear his proportional share of the expense to repair or rebuild it. Messrs. Baily, Hill and Evans consented, however, "to replace and properly attend to the putting in new Braces to the said Stack where they may now be wanting at their own Cost and Charge" as part of the initial repairs needed to be made, including their "undertaking to cover the Stack of said Furnace with a sufficient Roof to shelter the same from the approaching Winter."[263] The lessees were granted immediate access to the lands of the Andover Ironworks in order to prepare a stock of charcoal and to collect provisions for use in their operations.

On January 20, 1790, the owners were empowered to enter the lands and tenements of John Armstrong at Andover Forge and to distrain property equal in value to the sum of £360 or eight tons of good bar iron.[264] In response, John Armstrong agreed on March 30, 1790, to yield up the forge, dwelling houses, tenements, lands and mills on or before May 1.[265] He further agreed "that he will give up the Possession of the Mansion house and the use of the Kitchen to such Person as the Owners may wish to occupy the same, except the small Room now used as a Counting house, at any time after the present Week, it being expected that Mr. Cadwalader Evans with his family will arrive there this Week to take possession of the same."[266]

In April 1790, Cadwalader Evans occupied the mansion and kitchen house at Andover Forge (Waterloo).[267] Benjamin Chew and John Lardner authorized Baily, Hill and Evans to take possession of Andover Forge in May 1790.[268] Captain Cadwalader Evans and Abraham Baily commanded volunteer cavalry companies from Sussex County in 1794, during their three months' service in the Western Expedition to suppress the Whiskey Rebellion.[269] Cadwalader Evans's wife, Sarah Cox, died at Andover Forge in October 1794 and may be buried there.

Abraham Baily learned the rudiments of eighteenth-century surgery as a student of Dr. Nicholas Way, of Wilmington, Delaware, while attending the wounded after the battle of Brandywine Creek in 1777.[270] He then went to sea as surgeon on a privateer, visiting France during his cruise. After the war, he established a medical practice in his native West Bradford Township, Chester County, Pennsylvania, and married Phoebe Carpenter at Old Swedes Church, Wilmington, Delaware, on September 21, 1789.[271] He then engaged in the iron business with Humphrey Hill and Cadwalader Evans. According to a short biography, published in 1881, "They rented the Andover Works, in Sussex County, N.J., where they continued five or six years, but the enterprise resulted unfortunately."[272]

On June 25, 1795, the owners of the Andover lands were empowered to distrain the goods of their lessees for five years' unpaid rent, amounting to £636 8*s* 7*d*.[273] An inventory of the goods and chattels of Baily, Hill and Evans, taken December 28, 1795, showed slim pickings.[274] At Andover Forge they found two coal wagons, four horses, a harness for a wagon, four hundred bushels of coal in the coal house, two hundred weight of bar iron, five hundred weight of pig metal and forge tools. At Andover Furnace they found one horse, two cows, an ox, two tons of iron on the bank, twenty-five loads in the coal house, two coal wagons, two coal bodies, eight or ten old wagon wheels, furnace tools, two bad wagon bodies and blacksmith tools in the shop. At public vendue on June 12, 1796, the sale of goods distrained at Andover Furnace fetched £106 6*s* 1*d*.[275] Agent Samuel Fisher reported, "The Fences much out of order, the Meadow overflowed for want of drains to carry the water off, the Buildings all in want of Repairs. The gable end of the Mill [is] likely to fall in. Two farms, south of the Works, [are] in tolerable good repair; to the east, one small likewise [is] in good order; one to the north [is] in bad order. The big Meadow I did not examine, the reason why? It was covered with snow. The Forge Farm [at Waterloo] appears to have no Fence but looks like a Commons, the Forge [is] in bad order; the Mill in pretty good order. The Little Farm appears in bad order."[276] Scavengers were soon picking the bones. On June 27, 1797, Amos Grandin, of Schooleys Mountain, wrote to Benjamin Chew, informing him that Thomas Wills "has been with me to buy some brick and [iron] plate that is about the forge at Andover, which I refused to dispose of without your approbation."[277] While there was much material at the forge that could be sold, Grandin still expected to repair the forge shortly. In a letter, dated June 26, 1797, Thomas Wills asked to scavenge "part of the property, which is going to destruction as fast as it possibly can."[278] He "went in order to gather some brick, finding one chimney down and the rest on the brink of falling. I gathered up all the loose ones that were fallen out of the chimney." Following Mr. Grandin's suggestion, he sent his friend Isaac Hume to meet with the owners "to see if I cannot have some of the brick out of the chimneys. I will either pay you the money in hand or as many brick returned, if the work is repaired—which [ever] is suitable." He also offered to lease the Andover Forge Farm for the following season, if it was available.

On October 17, 1799, Benjamin Chew, John Lardner, William Allen and John Allen appointed Amos Grandin, of Morris County, their lawful attorney to receive all payments due from those who possessed any of the Andover tracts by reason of lease, written or verbal contract, by use or occupation.[279] And so the Andover Ironworks dwindled away. On March 6, 1802, Silas Dickerson, of Stanhope, wrote Benjamin Chew Jr. to say that he was still inclined to purchase the old Andover Forge Tract, despite Chew's earlier determination not to break up the Andover Ironworks and sell the forge tract separately.[280] The owners, however, conceived Dickerson's "offer less than the Value of the Property would justify under different Circumstances" but offered to sell all the Andover lands (excepting only the mine) for 33*s* 6*d* per acre, if the buyer would take "the Rough and Smooth."[281] Dickerson wrote again on October 13, 1806, wishing to know whether Chew and the other owners had changed their minds.[282] The owners, however, still placed a high value on the lands, noting, "There are two very valuable mines, very near to one another."[283] James Ludlum, of New York, thought otherwise and wrote Chew on August 31, 1807, saying, "In your letter you speak of Two very valuable Mines at or near the Furnace, which conveys an Idea that you consider the Estate very valuable on Account of those Mines, that however in our opinion, is not the fact, the Wood having been completely cut off by your former Tenants."[284]

The End

The absentee owners slowly came to recognize their predicament. On November 21, 1808, the General Assembly authorized a subdivision of the great land empire of the Andover Iron Company among the numerous heirs of the original owners, making it "lawful for any one of the justices of the Supreme Court of judicature of this state, on the application of any of the parties interested, to appoint three commissioners to make division of…the Andover Furnace and Forge Tracts, into such and so many shares as the same were originally held by the Andover Company or original owners thereof."[285] These commissioners would submit surveys for the proposed subdivision to the State Supreme Court for final approval. Judge Isaac Smith, a justice of the Supreme Court of Judicature, appointed David Frazer, Jacob Anderson and Benjamin Guild to survey a partition.

Benjamin Chew Sr. died January 20, 1810. A month later, his widow, Elizabeth Oswald Chew, conveyed her interest in the Andover Iron Company, which she inherited from Joseph Turner, to Benjamin Chew Jr., "for and in consideration of the natural love and affection, which she the said Elizabeth hath and beareth towards her said son."[286] On April 5, 1811, the various heirs of the owners of the Andover Iron Company agreed that Benjamin Guild and William Runkle should survey and divide the Andover lands. Of the 4,155.75-acre Andover Forge Lot (No. 7), 2,200 acres were accordingly deeded to the heirs of Henry Hill in compensation for their ownership of two and a half equal undivided sixteenth parts. This subdivision covered the southwestern slope of Allamuchy Mountain lying west and southwest of the

Andover Forge Farm (Waterloo), extending between Allamuchy Pond and the ridgeline on the Morris County side of the Musconetcong River.

On May 22, 1812, Elizabeth Chew and her son Benjamin sold their interest in eight parcels of land, comprising the Andover Furnace and Forge Tracts, to Dr. Isaac Ogden, of New Germantown, Hunterdon County, for $3,544 in cash and $6,448.10, secured by mortgage.[287] The sixty-acre Andover Mine Lot was excluded. On the same date, John Lardner, of Philadelphia, the surviving executor of Lynford Lardner's estate, also conveyed his quarter share to Isaac Ogden for $3,310.69 in cash and $5,571.50 secured by mortgage. Dr. Ogden and his partner, George C. Maxwell, a Flemington lawyer and congressman, thus acquired thirteen and a half equal undivided sixteenth shares in the Andover Iron Company's lands. The fifteen heirs of Henry Hill, Esq., late of Philadelphia, residing in Delaware, New Jersey, Pennsylvania and England, owned the remaining two and a half shares. On August 24, 1813, they released their two and a half equal and undivided sixteenth parts in the Andover lands to Dr. Isaac Ogden and George C. Maxwell.[288]

On May 25, 1812, Ogden and Maxwell sold the Andover Furnace Farm (Andover Borough) encompassing 707 acres, together with 15 acres at the outlet of Panther Pond, to Joseph Northrup for $15,290.[289] Despite the extinction of Andover Furnace, surviving accounts from January 1813 show William McKinney, James Brown, Lemuel De Camp, Andrew Sloughbower, Dr. Everitt and Mark Luse bringing buckwheat to Northrup's Mill for grinding. In 1816, Northrup converted the old furnace to a gristmill and the original gristmill to a store. On November 30, 1812, John Smith, of Roxbury, purchased the 282-acre Andover Forge Farm (Waterloo).[290] He soon enlarged his holdings there, paying $431 for 165.5 acres on the south side of Waterloo Lake to Abigail (Byram) Condict, of Morris Township, executrix, of Silas Condict, on March 13, 1813.[291] He also acquired an adjacent 50-acre tract located in Morris County from Edward Condict in 1816.

George C. Maxwell and Andrew Bartles sold parcels out of the old Folliott Tract, ranging between 18 and 242 acres, on July 21, 1810, to Joseph Desmund, Herbert Wire, Daniel Smith, Jacob Young, Robert Thompson, Lewis Thompson and Jacob Wick. The grantees were likely squatters on the land. Millwright Joseph Desmund (aka Demund) built a gristmill and distillery at the outlet of Allamuchy Pond on 42 acres, lying between the northwest edge of Allamuchy Pond and the Great Road from Newton to Hackettstown (Route 517), which he purchased of Ebenezer Willson on August 7, 1809, thus founding the village of Allamuchy.[292]

The Grist Mill Playhouse opened in the old Andover mill in 1950. *Photo by C.E. Engelbrecht, "Summer Stock,"* The Magazine Sussex, *July 1952.*

On May 10, 1812, Maxwell and Bartles conveyed 1,127 acres to attorney Nathaniel Saxton, of Flemington, for $3,920.[293] The boundary survey began at "a white oak by the great pond on Allamuchahocking Mountain." Nathaniel Saxton contributed his name to Saxton Falls. To consolidate their interests, the heirs of Henry Hill sold their portions of the Union Ironworks Tract in Hunterdon County and the 2,200 acres of the Andover Forge Lot in Sussex and Morris Counties to their attorney, James Vaux, of Philadelphia, for $10 in 1815. Vaux, acting as their trustee, appointed Nathaniel Saxton to manage these lands on April 2, 1816. Saxton was to take all lawful measures to prevent any further loss of "the timber and trees upon said premises." The agreement prohibited Saxton from giving leases "for longer than one year at any one time."[294]

As the Andover Ironworks fell into ruin, the memory of those who had lived, worked and died there began to fade. As late as 1854, two old burying grounds associated with Andover Furnace were situated on either side of the Sussex Railroad in the village of Andover. Most likely, this was a single plot, which the original road cut in two. According to a report published in 1881, "Traces of these grounds—used, of course, by the people employed at the iron-works—were visible until within a few years ago, but the headstones are now dislodged and the land broken to the plow."[295]

Writing to the *New Jersey Herald* in 1875, Seymour Smith, of Waterloo, described "an old Indian burying ground, in which the rough stone marks the final resting place of many a departed child of the forest." It is situated about a quarter of a mile east of the village on the shores of Waterloo Lake. The Smith family, owners of the land, "scrupulously respected" the burials and never plowed in their vicinity.[296] A promotional booklet for Lake Waterloo Estates, published about 1927, also reported, "An old Indian burying ground apparently was located in a rise of land which juts into the waters of what is now the main lake of a chain." A portion of this burying ground became an island in Waterloo Lake when the Mountain Ice Company excavated a channel to redirect the river's current along the northwestern bank and thus create a quiet pool of water, free of debris, for its ice harvests. Remnants of a dike, which once channeled the flow of water around the ice pond, are still discernible at the upper end of the lake. This ditch detached the peninsula-bearing part of the cemetery from the shore, thereby creating the smaller and more westerly of the two islands in Waterloo Lake.

This old cemetery at Waterloo reportedly encompasses about fifty interments, including the remains of the early inhabitants of Andover Forge and Byram Township. The promotional booklet for Lake Waterloo Estates claimed, "The health-giving atmosphere of the territory [around Waterloo] is attested by the fact that during the time of Washington's occupancy of headquarters in Morristown many of his French allied soldiers attacked by a small-pox were encamped in quarantine on a bluff overlooking the valley. A group of stones facing east marks the spot."[297] In 1980, Sandford Roy Smith (1887–1982), the son of Peter D. Smith, provided an eyewitness account of the excavation of graves in the Old Indian Burial Ground at Waterloo, which at this writing may underlie a simulated Lenape village, saying,

> *There is an area that has been known as the Indian burial ground, which is, of course, doubtful because Indians didn't bury below ground; they buried above ground.*[298] *With that in mind, approximately 60 years ago* [circa 1920] *a group of antiquarians legally came and opened three graves in the area. There are some fifty graves that are marked there, but three that were opened—these men knew what they were doing and it was judged that one was Irish, one was French and one was an Indian. The graves were all eight feet deep to the top of the coffin and the coffin was of the antique shape, narrow at the top and long to the feet, wide at the shoulders and had been made of hand-hewn planks and hand-wrought nails used to seal them up. In these graves were found several artifacts such as brass buttons, a pipe, and the quantities of hair still visible. These graves were closed subsequently and they are still there the same as they were in early times.*[299]

The Old Andover Gristmill (Furnace) in 1980. *Photograph by Kevin Wright.*

Roy reportedly collected pewter buttons and a clay pipe with a deer-horn stem, which had flowers painted on the pipe-bowl. Another amateur archaeologist explored this old burial ground in 1945, when Waterloo Road was built on its present route, unearthing a casket containing a skeleton that reportedly still displayed a full head of red hair.

Notes

Place in History

1. Lesley, *Iron Manufacturers' Guide*, 339.
2. Kitchell, *Third Annual Report*, 10; Lesley, *Iron Manufacturers' Guide*, 416.
3. Cook, *Geological Survey of New Jersey*, 11.
4. Lesley, *Iron Manufacturers' Guide,* 428–29.
5. James Thatcher Hodge (1816–1871), a native of Plymouth, Massachusetts, graduated from Harvard in 1836 and became active in numerous field surveys throughout the United States. A respected member of the Association of American Geologists, he was professor of geology at Cooper Institute in New York. Tenney, *Mining Magazine*, 315.
6. Lesley, *Iron Manufacturers' Guide,* 428–29.

Discovery

7. Sims and Leonard, *Geology of Andover Mining District*, 26; Bayley, *Mining in New Jersey*, 79.
8. It was erroneously listed as "No. 75. 1250 Acres at the Head of Pequase {*sic*} River, near Major Woolverton" in a list of sundry tracts of land that Thomas and Richard Penn offered for sale on March 10, 1755. Nelson, *NJ Archives, Extracts from American Newspapers, 1751–1755*, 467.
9. Kitchell, *Third Annual Report*, 18.

10. See Hermelin, *Report About the Mines*, 55–56; Tiemann, *Iron and Steel*, 135; Overman, *Manufacture of Iron*, 247–48.
11. Tiemann, *Iron and Steel*, 136.
12. Overman, *Manufacture of Iron*, 255–56.
13. Tiemann, *Iron and Steel*, 139.
14. Kitchell, *Third Annual Report*, 19. Some Highland bloomeries produced anchors, which found a convenient market in New York Harbor.
15. Snell, *Sussex and Warren Counties*, 431. According to the *Annual Statistical Report of the American Iron & Steel Association*, published in 1908, no bloomeries producing blooms directly from iron ore had operated in the United States since 1901, in which year bloomeries produced 2,310 gross tons, against 4,292 tons in 1900 and 3,142 tons in 1899. The last bloomeries or Catalan forges apparently operated in the South.
16. Those existing when the act passed were allowed to continue, namely, the Trenton Steel Works, Samuel Ogden's slitting mill near Boonton and Allen and Turner's rolling-and-slitting mill at the Union Iron Works.
17. Hermelin, *Report About the Mines*, 15.
18. As their deed of purchase reportedly mentioned an existing forge, it appears that ore mined at High Bridge was worked at a bloomery for some time prior to construction of Union Furnace on Spruce Run.
19. A specimen of pig iron from the old Andover charcoal furnace assayed 89.440 percent iron, 5.750 percent manganese, 4.529 percent combined carbon, 0.101 percent graphite, 0.060 percent phosphorus, 0.030 percent silicon and 0.003 percent sulfur. Bayley, *Mining in New Jersey*, 82. See also Tiemann, *Iron and Steel*, 258.
20. Honeyman, *Extracts from American Newspapers, 1775*, 55.
21. William Allen was born August 5, 1704, the son of William and Margaret Budd Allen. He married Margaret Hamilton, daughter of Andrew Hamilton, Speaker of the Provincial Assembly, in February 1733. They had six children: John; Andrew; William; James; Ann, wife of Governor John Penn; and Margaret, wife of James DeLancey. As head of the Gentlemen's Party, Judge Allen defended the Penns' proprietary overlordship against Benjamin Franklin's Country Party of Germans and Quakers. Benjamin Chew succeeded him as Pennsylvania's supreme justice in 1774. William Allen died September 6, 1780.
22. Joseph Turner was born in Andover, Hampshire, England, on May 2, 1701. A sea captain, he settled in Philadelphia in 1714, where he engaged in shipbuilding with Clement Plumstead, establishing a merchant fleet and outfitting privateers. Joseph Turner became a Philadelphia councilman in 1729. He died on July 25, 1783, at eighty-two years of age.

23. "Deposition of John Hackett of the Province of New Jersey—relating to a riot committed on the estate of Messrs. Allen and Turner," in Whitehead, NJ *Archives, Vol. VII, Part of the Administration of Governor Jonathan Belcher, 1746–1751*, 377–79.
24. Elizabeth Reading Hackett died on March 17, 1781, aged fifty-three years, at Roxbury, Morris County. See the *New Jersey Journal*, March 28, 1781, in Scott, NJ *Archives, Volume V, Extracts from American Newspapers Relating to New Jersey, October, 1780–July, 1782.*

ANDOVER FURNACE

25. Lynford Lardner was born near London, England, on July 18, 1715. He apprenticed with his brother-in-law, Richard Penn, youngest son of William Penn, in the woolen industry but moved to Pennsylvania in 1740, where he was employed in the Land Office, serving as receiver-general in the collection of quitrents due the Proprietors from 1741 to 1753. As a large landowner in his own right, he invested in iron manufactures in Berks and Lancaster Counties, most notably at Windsor Forge in Lancaster County. He married Elizabeth Branson, daughter of a wealthy Philadelphia merchant, in 1749. They had seven offspring before she died in 1760. In 1766, he married Katherine, daughter of Philadelphia mayor Thomas Lawrence. Lardner was appointed to the Governor's Council in 1755. His nephew, John Penn, son of Richard and Hannah Penn, became Pennsylvania's last proprietary governor in 1763. Lynford Lardner died October 6, 1774.
26. This tract was originally surveyed for the wife of John Evans, Esq., of Hunterdon County, in 1715. William Moore's sister, Rebecca, married John Evans, lieutenant governor of Pennsylvania, in 1708. He and his wife moved to England shortly after their marriage.
27. It has been called Piggot's Pond, Andover Long Pond, Andover Great Pond, Struble's Pond, Slaters Lake and New Wawayanda Lake.
28. In 1810, the Hall's Pond Lot was known as the Wilson farm.
29. *Sussex County Road Book A*, 2.
30. Upon dispersal of the Andover Iron Company's lands, this tract was subdivided into five farms, ranging in size from 58.5 to 262.5 acres.
31. Prior to the fourteenth century, molten iron was only produced as an accidental byproduct of the Catalan forge, an ancient bloomery process. See Tiemann, *Iron and Steel*, xii. A charcoal blast furnace produced pig iron in Westphalia by 1311, and another operated in Sussex, England, before 1360. For dimensions, see Hermelin, *Report About the Mines*, 59.

32. From the German *boschung*, meaning "slope."
33. Nelson, *NJ Archives, Extracts from American Newspapers, 1756–1761*, 476.
34. Walker, *Burd Papers*, 44.
35. Tiemann, *Iron and Steel*, 37.
36. Hermelin, *Report About the Mines*, 59–60.
37. Kitchell, *Third Annual Report*, 22.
38. Ibid.
39. Hermelin, *Report About the Mines*, 60.
40. Moldenke, *Principles of Iron Founding*, 68–70; Tiemann, *Iron and Steel*, xi–xii.
41. Slag was also called cinder or scoria.
42. Tiemann, *Iron and Steel*, 35.
43. Moldenke, *Principles of Iron Founding*, 39.
44. Ibid., 40–44.
45. Ibid., 44–47.
46. Ibid., 47–49.
47. Ibid., 49–51.
48. Tiemann, *Iron and Steel*, xii.
49. Description of these appendages can be found in three surviving inventories, namely: "An Inventory of the moveable effects at Andover Furnace belonging to the Owners of said works," compiled by Ephraim Darby and Nathaniel Pettit on June 11, 1767; a second inventory and appraisal, made by Nathaniel Pettit, William Crooks and Amos Pettit "at the time of Mr. Lardner's renting his Share to Mr. Allen, Mr. Turner and Mr. Chew, being the 15th of May 1772"; and a third compiled by ironmaster John Jacob Faesch, with assistance from Charles Craig, on June 26, 1782. Chew Family Papers.
50. The *Sussex Register* reported on June 8, 1876, "The old dam at the mill in Andover was built in 1760 and yet the pointing is as perfect as if it just had been completed."
51. Walker, *Burd Papers*, 43–44.
52. Overman, *Manufacture of Iron*, 393.
53. Hans Lobsayer reportedly invented the blowing tub about 1550 at Nuremburg, Germany.
54. Schultz, *Antique Iron Works*, 6.
55. Ibid.
56. Ibid.
57. Ibid.
58. Tiemann, *Iron and Steel*, 36.
59. Account of Isaac Hall to Andover Furnace, Chew Family Papers.
60. Hermelin, *Report About the Mines*, 60–61.
61. Tiemann, *Iron and Steel*, 299.

62. Goodrich, *New Family Encyclopedia*, 198–99.

63. Hermelin, *Report About the Mines,* 47–51.

64. Walker, *Burd Papers*, 43–44.

65. Charcoal made from good hardwood is 84 to 89 percent fixed carbon, 4 to 5.5 percent volatile matter, 4.5 percent moisture, 1 to 2 percent ash, .05 percent sulfur and .05 percent phosphorus. Moldenke, *Principles of Iron Founding*, 240–41; Tiemann, *Iron and Steel,* 74; see also Kemper, *American Charcoal Making*, 1941.

66. Tiemann, *Iron and Steel,* 74

67. Overman, *Manufacture of Iron*, 84. According to Overman, "A cord contains 128 square feet; that is, the billets must be four feet long, and the cord four feet high and eight feet long. A great deal of deception is practiced by workmen, who need close watching."

68. Hermelin, *Report About the Mines,* 48.

69. Schultz, *Antique Iron Works,* 1–3.

70. Overman, *Manufacture of Iron*, 85.

71. Hermelin, *Report About the Mines,* 48.

72. *East Jersey Deed Book S-10*, 290.

73. Hermelin, *Report About the Mines,* 47–48; Overman, *Manufacture of Iron*, 84.

74. Schultz, *Antique Iron Works,* 1–3. Overman (116) claimed, "One hundred pounds of wood in kilns produce on an average, twenty pounds of charcoal."

75. Hermelin, *Report About the Mines,* 48.

76. "Re: Charcoal: Estimate of usage," Tue, 1 Jun 1999 16:21:51 -0400, arch-metals archives - June 1999 - Estimates of usage, http://www.mailbase.ac.uk/lists-a-e/arch-metals/1999-06/0007.html.

77. Inventory and Appraisement of the Stock and Utensils &c remaining at and belonging to Andover Furnace at the time Mr. Lardner's renting his Share to Mr. Allen, Mr. Turner and Mr. Chew, being the 15th of May 1772, Chew Family Papers.

78. Nelson, *NJ Archives, Extracts from American Newspapers, 1772–1773*, 404, 440.

79. Nelson, *NJ Archives, Extracts from American Newspapers, 1770–1771*, 271–73. For description of the Andover mine in 1783, see Hermelin, *Report About the Mines,* 24.

80. Inventory and Appraisement of the Stock and Utensils &c remaining at and belonging to Andover Furnace at the time Mr. Lardner's renting his Share to Mr. Allen, Mr. Turner and Mr. Chew, being the 15th of May 1772, Chew Family Papers.

81. Hermelin, *Report About the Mines,* 26.

82. Tiemann, *Iron and Steel,* xii.

83. Water struck the wheel at about three-quarters of its height or "high on its breast." Inventory made by John Jacob Faesch, with assistance from Charles Craig, on June 26, 1782, Chew Family Papers.

84. Nelson, *NJ Archives, Extracts from American Newspapers, 1770–1771*, 271–73.
85. Inventory and Appraisement of the Stock and Utensils &c remaining at and belonging to Andover Furnace at the time Mr. Lardner's renting his Share to Mr. Allen, Mr. Turner and Mr. Chew, being the 15th of May 1772, Chew Family Papers.
86. Ibid; Inventory of the moveable effects at Andover Furnace belonging to the Owners of said Works, appraised by Nathaniel Pettit, and Ephraim Darby, Esquires, June 11, 1767, Chew Family Papers. The Hopkins-Allen *Map of Sussex County, New Jersey* (1860) shows a blacksmith shop on the southeast side of the triangular green formed by the bifurcation of the Andover-Mohawk Road as it intersects Route 206. It, however, may postdate the Andover Iron Company. It is possible that a stone building, standing south of the original furnace, was the original blacksmith shop. It is known only from early twentieth-century postcard views.
87. Inventory of the moveable effects at Andover Furnace belonging to the Owners of said Works, appraised by Nathaniel Pettit, and Ephraim Darby, Esquires, June 11, 1767, Chew Family Papers.
88. Nelson, *NJ Archives, Extracts from American Newspapers, 1770–1771*, 271.
89. Inventory and Appraisement of the Stock and Utensils &c remaining at and belonging to Andover Furnace at the time Mr. Lardner's renting his Share to Mr. Allen, Mr. Turner and Mr. Chew, being the 15th of May 1772; Inventory of the moveable effects at Andover Furnace belonging to the Owners of said Works, appraised by Nathaniel Pettit, and Ephraim Darby, Esquires, June 11, 1767, Chew Family Papers.
90. Inventory of the moveable effects at Andover Furnace belonging to the Owners of said Works, appraised by Nathaniel Pettit, and Ephraim Darby, Esquires, June 11, 1767, Chew Family Papers.
91. Inventory and Appraisement of the Stock and Utensils &c remaining at and belonging to Andover Furnace at the time Mr. Lardner's renting his Share to Mr. Allen, Mr. Turner and Mr. Chew, being the 15th of May 1772, Chew Family Papers.
92. Hermelin, *Report About the Mines,* 61.
93. Nelson, *NJ Archives, Extracts from American Newspapers, 1756–1761*, 544–45.
94. Inventory of the moveable effects at Andover Furnace belonging to the Owners of said Works, appraised by Nathaniel Pettit, and Ephraim Darby, Esquires, June 8, 1767; Inventory and Appraisement of the Stock and Utensils &c remaining at and belonging to Andover Furnace at the time Mr. Lardner's renting his Share to Mr. Allen, Mr. Turner and Mr. Chew, being the 15th of May 1772, Chew Family Papers.

95. Inventory of the moveable effects at Andover Furnace belonging to the Owners of said Works, appraised by Nathaniel Pettit, and Ephraim Darby, Esquires, June 8, 1767; Inventory and Appraisement of the Stock and Utensils &c remaining at and belonging to Andover Furnace at the time Mr. Lardner's renting his Share to Mr. Allen, Mr. Turner and Mr. Chew, being the 15th of May 1772, Chew Family Papers.
96. Hermelin, *Report About the Mines*, 22.
97. "Life at Andover Iron Works in 1763," *Sussex Register*, October 12, 1893.
98. The initials "J.S.W." stand for John Sylvester White.
99. "Andover Items," *Sussex Register*, February 24, 1876.
100. Nelson, *NJ Archives, Extracts from American Newspapers, 1770–1771*, 271.
101. Ibid., 271–73.
102. Inventory made by John Jacob Faesch, with assistance from Charles Craig, on June 26, 1782, Chew Family Papers.
103. Inventory and Appraisement of the Stock and Utensils &c remaining at and belonging to Andover Furnace at the time Mr. Lardner's renting his Share to Mr. Allen, Mr. Turner and Mr. Chew, being the 15th of May 1772, Chew Family Papers.
104. Nelson, *NJ Archives, Extracts from American Newspapers, 1762–1765*, 571.
105. Ibid.
106. *East Jersey Deeds, S-10*, 44. A tract of 62.89 acres, acquired by the heirs of John Johnston on June 11, 1791, was "situate on both sides of a northerly branch of the Pequest called the Andover Saw Mill Brook." The Andover Saw Mill probably stood near the present outlet of Lake Lenape.
107. Inventory made by John Jacob Faesch with assistance from Charles Craig, on June 26, 1782, Chew Family Papers.

ANDOVER FORGE

108. Schultz, *Antique Iron Works*, 9.
109. Ibid.
110. Ibid.
111. Ibid., 11.
112. "Whisperings from Waterloo," *New Jersey Herald*, May 5, 1875.
113. This tract was included in a survey of 168 acres made out to the London Company in 1738. After the Quintipartite Line was drawn in 1743, nine-tenths of it fell within East Jersey. In compensation, the West Jersey Proprietors issued a new warrant to the London Company on February 4, 1762, allowing it to relocate 159 out of the 168 acres elsewhere.

114. Nelson, *NJ Archives, Extracts from American Newspapers, 1756–1761*, 512.
115. The Folliott Tract extended from Allamuchy Pond and the middle of what is now Saxton Lake down the valley to Stephens State Park and Hackettstown.
116. John and Dorothy Hunt, of London, conveyed the Garbut & Breckon Tract to Allen, Turner, Lardner and Hackett on January 29, 1763.
117. This deed describes Budd Lake as "the little pond being the head of the South Branch of the Raritan." According to an advertisement in the *Pennsylvania Gazette* on October 21, 1762, the George Eyre Tract comprised "about 943 Acres, the greatest part of which is well timbered, and about 90 Acres of the Tract is cleared, and under good Fence, with a good Log House on it; there is a constant Stream or Brook of Water, running thro' the said Tract, very suitable to erect a Grist-mill or Forge on, and a great Part will make good Meadow."
118. Though it remains unverified, S. Roy Smith (1887–1982) claimed, "The original river bed ran along the southerly side of the [Waterloo] lake, running between the present island and where the ice houses were [on the south shore]. The southerly side of the [Morris Canal] Lock Pond is also the deepest, marking the original channel of the river." Roy Smith also claimed that the first forge at Waterloo was built down at the end of the dam on the Morris County side of the river. All the evidence, including contemporary maps, shows Andover Forge on the Sussex County side of the river.
119. Lesley, *Iron Manufacturers' Guide,* 429.
120. Webb, *Sussex County, N.J. Historical Directory*, 23.
121. "Whisperings from Waterloo," *New Jersey Herald,* May 5, 1875.
122. Inventory made by John Jacob Faesch, with assistance from Charles Craig, on June 26, 1782, Chew Family Papers (Collection 2050), the Historical Society of Pennsylvania.
123. Today it widens into the millpond and passes underneath the gristmill, returning to the river below the canal lock behind the blacksmith shop. The Lock Pond didn't exist in the eighteenth century and may drown evidence of earlier mills and raceways. It seems the original dam was situated where nineteenth-century bridge abutments mark the spot where an iron bridge later carried Taylor Road across the Musconetcong River near the Waterloo grist- and sawmills.
124. Hermelin, *Report About the Mines,* 64, 69.
125. Ibid.
126. Overman, *Manufacture of Iron*, 256–59.
127. Inventory made by John Jacob Faesch with assistance from Charles Craig, on June 26, 1782, Chew Family Papers.
128. Ibid.
129. Ibid.

130. Ibid.
131. Ibid.
132. Walker, *Burd Papers*, 43.
133. Schultz, *Antique Iron Works,* 9–10.
134. Ibid., 9.
135. Tiemann, *Iron and Steel,* 75.
136. Ibid.
137. Hermelin states that a ton was called a hundredweight (cwt.) and that twenty-two hundredweights of anconies were delivered out of twenty-eight hundredweights of pig iron. Twenty-two hundredweights of anconies made twenty hundredweights of bar iron. See Hermelin, *Report About the Mines,* 20, 64.
138. Inventory made by John Jacob Faesch with assistance from Charles Craig, on June 26, 1782, Chew Family Papers.
139. Ibid.
140. Ibid.
141. Ibid.
142. Ibid.
143. Ibid.
144. Ibid.
145. Ibid. It is possible the blacksmith shop and magazine underlies the double stone tenant house standing opposite the Smith Brothers' Gristmill.
146. Ibid.
147. *Sussex County Road Book A*, 11. A rod is sixteen and a half feet; hence, a four-rod road is sixty-six feet wide, sixty-six feet being the equivalent of the standard one-hundred-link surveyor's "chain." Sixty-six feet is still the standard right-of-way for public highways.
148. Ibid.
149. "Whisperings from Waterloo," *New Jersey Herald*, May 5, 1875.
150. Nelson, *NJ Archives, Extracts from American Newspapers, 1762–1765,* 293.
151. Ibid., 427.
152. Walker, *Burd Papers*, 69.
153. Ibid., 45.
154. "Selections from the Correspondence of William Alexander, Earl of Stirling," *Proceedings of the New Jersey Historical Society, Vol. VII, 1853–1854, No. 2*, 112.
155. Nelson, *NJ Archives, Extracts from American Newspapers, 1766–1767*, 243.
156. *NJ Archives, Calendar of New Jersey Wills, Administrations, etc.*, Vol. IV—1761–1770, 168.
157. Articles of Agreement between Archibald Stewart and Joseph Reid, Administrators of the estate of John Hackett, deceased, and Benjamin Chew, August 24, 1767, Chew Family Papers.

158. Ibid.
159. Ibid.
160. Letter of Benjamin Chew to Joseph Reid, October 26, 1767, Chew Family Papers.
161. Walker, *Burd Papers*, 73–74.
162. Nelson and Honeyman, *NJ Archives, Extracts from American Newspapers, 1773–1774*, 541.

BLISTER STEEL

163. It was called a "cementation" process because charcoal layers resembled "cement" joints between stacked iron bars; see Tiemann, *Iron and Steel,* 70–71.
164. Acrelius, *A History of New Sweden*, 167.
165. Hall, *History of the Presbyterian Church*, 110; *American Weekly Mercury*, September 5, 1734; *American Weekly Mercury*, No. 168, September 12–19, 1734.
166. Hall, *History of the Presbyterian Church*, 111; *Pennsylvania Gazette*, August 15, 1745; Whitehead, *NJ Archives, Part of the Administration of Governor Jonathan Belcher, 1746–1751*, 558–59.
167. Nelson, *NJ Archives, Extracts from American Newspapers, 1772–1773*, 458.
168. *Journals of the Continental Congress*, 14, 515.
169. Letter from John Hackett, of the Andover Ironworks, to Whitehead Humphries, steel manufacturer, of Philadelphia, June 21, 1766, Chew Family Papers.

DEPRESSION

170. Kimball and Quinn, "William Allen–Benjamin Chew Correspondence, 1763–1764," 224.
171. Hermelin, *Report About the Mines,* 15.
172. As of May 30, 1767, Joseph Turner's accounts indicate that Andover Furnace accrued debts to the amount of £16,143.15*s*.9*d*, while debts due to Andover Forge amounted to £3,535.10*s*.27*d*. See Chew Family Papers.
173. Nelson, *NJ Archives, Extracts from American Newspapers, 1766–1767,* 286.
174. William Allen to Thomas Penn, October 8, 1767, *Penn Official Correspondence*, MSS, Vol. X, 116, quoted from Kistler, "William Allen, Founder of Allentown," *Proceedings*, 14.
175. William Allen to Thomas Penn, September 23, 1768, *Penn Official Correspondence*, 170, quoted from Kistler, 14, 15.
176. Walker, *Burd Papers*, 77.

177. Brothers Archibald, William and James Stewart came from Ballantoy, County Antrim, Ireland, and settled in the vicinity of Hackettstown, New Jersey. Archibald was born August 18, 1737. He married Margaret (Phebe) Helms in August 1772 and died in May 1795, aged fifty-nine, at Springdale, Andover Township, Sussex County, New Jersey. On April 8, 1797, Phebe Stewart, his widow, sold 354 acres of woodland, known as Springdale Farm, to William Helm of Hackettstown; 266 acres in Independence Township, Sussex County; and land in Hackettstown "where Phebe now lives" (*Sussex County Deed Book C*, 205). William Stewart was born in 1739 and died in Hackettstown on February 17, 1810. By his wife, Frances Sherrod, he had six children: Jane Chitester; Samuel, born December 28, 1768; John, born April 14, 1770, who married Sarah Bird and died August 1, 1836; James, born 1772, who married Elizabeth Culver; Sarah (Helms); and Frances, born February 20, 1780, who married John Bird and died August 18, 1849. James and Alexander Stewart advertised the sale of paint at Hackettstown in 1778. On September 5, 1778, James Stewart was named a Loyalist.

178. Nelson, *NJ Archives, Extracts from American Newspapers, 1770–1771*, 241–42.

179. Ibid., 271–73.

180. *Sussex County Deed Book A*, 41.

181. Lease from Lynford Lardner to Allen and Turner, May 15, 1772, Inventory and Appraisement of the Stock and Utensils &c remaining at and belonging to Andover Furnace at the time Mr. Lardner's renting his Share to Mr. Allen, Mr. Turner and Mr. Chew, being the 15th of May 1772, Chew Family Papers.

182. Stryker, *NJ Archives, Extracts from American Newspapers, 1776–1777*, 388–89.

183. Nelson and Honeyman, *NJ Archives, Extracts from American Newspapers, 1773–1774*, 405.

184. Chew Family Papers.

Revolutionary War

185. Articles of Agreement between William Allen and Joseph Turner and John Patton, Esquire, of Bucks County, Pennsylvania, September 20, 1777, Chew Family Papers. Born in Sligo, Ireland, in 1745, John Patton came to Philadelphia. After marrying Brigitte Huling Bird, the widow of ironmaster William Bird in 1762, he became part owner of Roxborough Furnace, which he renamed Berkshire Furnace, in Heidelberg Township, Berks County. He served as colonel of the Fifth Battalion of Associators from Berks County, Pennsylvania, during the Revolution. He died in Centre County, Pennsylvania, in 1804.

186. Ibid.

187. Declaration of Col. Benjamin Flower, Philadelphia, August 19, 1778, Letter and Reports, 1781–88, from John Pierce, Paymaster General and Commissioner for Army Accounts, and Records, Relating to Investigations of Treasury Offices, 1780–81, National Archives Catalog ID 1938489, NARA M247, Item 62, Roll 76, *Papers of the Continental Congress*, 639–50. Congress appointed Benjamin Flower the commissary of military stores for the armies of the United States on July 16, 1778. Six months later, on January 16, 1777, General George Washington appointed him the commissary general for military stores with the rank of lieutenant colonel in the artillery of the United States. Washington instructed Flower "relative to sundry Works to be Established at Carlisle, and for raising a Corps of Artillery Artificers and also to procure a large quantity of Military Stores and to establish the necessary Officers to execute the different duties necessary for the supplies of the Army, &c."

188. Cook, *Geology of New Jersey*, 640–41n; Ford, *Journals of the Continental Congress 1774–1789*, 56.

189. Cook, *Geology of New Jersey*, 641n.

190. Prince and Ryan, *The Papers of William Livingston, Vol. 2: July 1777–December 1778*, 222–23.

191. *Correspondence of Executive of New Jersey, From 1776 to 1786*, 113–15.

192. Ibid., 115.

193. Thomas Mayberry was born in St. John-in-Bedwardine, Worcestershire, England, on August 22, 1738. He was a son of Thomas Mayberry, ironmaster of Powick Forge, and his wife, Mary. He immigrated to America around 1763 and married Cynthia Lanning of Middlesex County, New Jersey, on May 11, 1774. He operated the ironworks at Mount Holly, New Jersey, from 1775 to June 1778, when British troops destroyed it. After the death of his first wife, Thomas Mayberry, of Mount Holly, married Polly Spring, of Maryland, on February 3, 1780, but on March 12, 1781, he married Mary Sincleair. Thomas Mayberry moved to Hawkins, North Carolina, before December 26, 1791, and then to Charleston, South Carolina, before 1800, where he operated a boardinghouse. He died in Charleston, South Carolina, on March 6, 1819. See Avery, *The American Monthly Magazine*, 26 (January–June 1905): 141.

194. *NJ Archives, Selections from the Correspondence of the Executive of New Jersey from 1776 to 1786*, 114.

195. Ibid.

196. Prince and Ryan, *The Papers of William Livingston, Vol. 2: July 1777–December 1778*, 358–59.

197. *Acts of the General Assembly, June 17–22, 1778*, 85–86. While the names of the commissioners remain uncertain, they were possibly Mark Thompson, Samuel

Breckinridge and John Bowne. Bowne or possibly his heirs sued Mark Thompson and Samuel Breckinridge in 1788 for compensation they promised him in 1784 for managing the Andover Ironworks. Bowne won a judgment of $400 from Thompson in 1790.

198. Maybery, Thomas, Account with Bd. of War, 1778. 3 p. Settled by Benjamin Flower, witnessed by William Thorne. December 31, 779. Roll 55, Record Group 360, National Archives Catalog ID 1938489, "Petitions Addressed to Congress 1775–89," *Papers of the Continental Congress*, Vol. 5, M-N, 184.

199. Record Group 93, War Department Collection of Revolutionary War Records, 1709–1936 (ARC Identifier 604855), 035389. See also Ford, *Journals of the Continental Congress 1774–1789*, 815.

200. NARA M853, Roll 0040, National Archives Catalog ID 607212, Orders for Pay by the Commissary General of Military Stores Department, 03/21/1780–10/02/1780, Vol. 128, "June 24, 1780, Order in favor of Col. Thomas Maybury & Co. for £986.18," 56. Taunton Forge was six miles from Ancocus Landing.

201. The Convention troops were the surrendered army of General Burgoyne at Saratoga who were marched to places of safety in the interior of Pennsylvania and elsewhere. Prince and Ryan, *The Papers of William Livingston, Vol. 3: January 1779–June 1780,* 17.

202. *Papers of the Continental Congress*, No. 147, III, folio 153; March 30, 1779, *Journals of the Continental Congress, Volume XIII*, 406.

203. NARA M853, Roll 0040, National Archives Catalog ID 607212, Orders for Pay by the Commissary General of Military Stores Department, 03/21/1780–10/02/1780, Vol. 128, "June 24, 1780, Order in favor of Col. Thomas Maybury & Co. for £986.18," 56.

204. "Estimate of the Quantity & Value in Specie of the Articles to be purchased by Contract and Materials necessary to employ the Men belonging to the Regiment of Artillery artificers for procuring Supplies for the Armies of the United States for the ensuing Campaign," *Papers of the Continental Congress*, Record Group 360, NARA M247, Item Number 147, Reports of the Board of War and Ordnance 1776–81, Vol. 6, December 1780–April 1781, 17.

205. James Morgan is likely the ironmaster associated with Durham Furnace in Bucks County, Pennsylvania. He died in 1782.

206. Hunt, *Journals of the Continental Congress 1774–1779*, 259, and *Papers of the Continental Congress*, No. 136, IV, folio 147.

207. *Journals of the Continental Congress 1774–1779, Volume XVI, 1780 January 1–May 5*, 272; *Papers of the Continental Congress*, No. 42, V, folio 179, NARA M247, Item 136, Reports of the Board of Treasury, 1776–81, Vol. 3, 1779, Vol. No. 4, Roll 147, 147.

208. Letters from the Board of War and Ordnance, 1780–1781, 1778. NARA M247, The correspondence, journals, committee reports, and records of the Continental Congress (1774–1789), Record Group 360, Item 148, Roll 161, National Archives Catalog ID 1938489, *Papers of the Continental Congress*, Vol. 1, 66–71.
209. *Journals of the Continental Congress 1774–1779, Volume XVI, 1780 January 1–May 5*, 280; *Papers of the Continental Congress*, No. 148, I, folio 67.
210. *Journals of the Continental Congress 1774–1779, Volume XVI, 1780 January 1–May 5*, 319–20.
211. *Papers of the Continental Congress*, National Archives Catalog ID 1938489, Record Group 360, NARA M247, Item 42, Vol. 5 M–N, Roll 55, 187.
212. Estimate of ammunition by requisition of H. Knox, Letters and other records of the Committee to Headquarters (Philip Schuyler, John Matthews and Nathaniel Peabody) appointed to confer with the Commander in Chief, 1780, *Papers of the Continental Congress*, Record Group 360, NARA 247, Roll 46, Item 39, Vol. 3, 458.
213. Numbered Record Books Concerning Military Operations and Service, Pay and Settlement of Accounts, and Supplies in the War Department Collection of Revolutionary War Records, NARA M853, Roll 0037, National Archives ID 607153, Military Stores Received and Delivered at Philadelphia, March 1780–September 1784, Volume 133, "Journal of Military Stores, March 1780–March 1781," 6.
214. Scott, *Newspaper Extracts, October 1780–July 1782*, 104.
215. Inventory made by John Jacob Faesch with assistance from Charles Craig, on June 26, 1782, Chew Family Papers. Charles Craig may be the captain who served in Washington's military intelligence service.
216. Numbered Record Book Concerning Military Operations and Service, Pay and Settlement of Accounts, and Supplies in the War Department Collection of Revolutionary War Records, Record Group 93, NARA M853, Roll 0033, National Archives ID 604855, Volume 92, Letters Sent by Samuel Hodgdon, Assistant Commissary General of Military Stores, Commissary of Military Stores, and Assistant Quartermaster, January 1, 1781–May 24, 1784, July 1778– May 1784, Letter from Samuel Hodgdon, Philadelphia, 27 March 1782, to Mr. Thomas Potts, 170.
217. Various Books of Samuel Hodgdon, Commissary General of Military Stores, Numbered Record Books Concerning Military Operations and Service, Pay and Settlement of Accounts, and Supplies in the War Department Collection of Revolutionary War Records, Record Group 93, NARA M853, Roll 0036, Catalog ID 607363, "Account Book of Military Stores, 1780 to 1792," 6.
218. Letters Sent by Samuel Hodgdon, Assistant Commissary General of Military Stores, Commissary of Military Stores, and Assistant Quartermaster, January 1, 1781–May 24, 1784, Vol. 92, Numbered Records Books Concerning Military

Operations and Service, Pay and Settlement of Accounts, and Supplies in the War Department Collection of Revolutionary War Records, NARA M853, Roll 033, National Archives ID 604855, 88–89.

219. Ibid., 203–4.

220. Ibid.

221. Ibid., 203.

222. Lease of Andover Ironworks to Archibald Stewart from Joseph Turner and rest of owners, November 23, 1782, Chew Family Papers.

223. Various account entries for the Andover Company, Chew Family Papers.

224. Hermelin, *Report About the Mines,* 16.

225. Ibid.

226. "Historical Notes on Forests," *Annual Report of the State Geologist for the Year 1899, Report on Forests*, 2–3.

PEACE RETURNS

227. Hermelin, *Report About the Mines,* 65–66.

228. Ibid., 69.

229. Ibid., 67–68.

230. *Pennsylvania Gazette*, December 7, 1785; January 4, 1786; August 9, 1786; *Minutes of the Supreme Executive Council of Pennsylvania, Vol. XV*, 11.

231. Mulholland, *History of Metals*, 147.

232. Coxe, *Lord Sheffield's Observations*, 25.

233. Letter from Archibald Stewart to Kuhn & Risberg, February 27, 1784, Chew Family Papers.

234. Letter from Archibald Stewart to Kuhn and Risberg, June 16, 1784, Chew Family Papers.

235. Letter from Benjamin Chew Jr. to Benjamin Chew Sr., December 18, 1784, Chew Family Papers.

236. The road ran about 230 feet northwest from the southeast corner of the coal house, thence southeast, 264 feet, and southwest, 198 feet, to the edge of the Musconetcong River. It is difficult to imagine this configuration today. There was a public bridge at this location until a large truck collapsed the last one around 1967.

237. Chew Family Papers.

238. *Independent Journal,* June 15, 1785 in Wilson, *Notices from New Jersey Newspapers,* 444.

239. Captain John Stotesbury and Hugh Hughes, Esquire, Account with Archibald Stewart; Letter from Archibald Stewart, of Hackettstown, to Kuhn & Risberg, April 27, 1786, Chew Family Papers.

240. Letter from Benjamin Chew to Robert Taylor, August 20, 1786, Chew Family Papers.
241. Ibid.
242. Ibid.
243. Letter from Benjamin Chew to Robert Taylor, February 1787, Chew Family Papers.
244. Letter from Benjamin Chew to Robert Taylor, August 20, 1786, Chew Family Papers.
245. Accounts, Chew Family Papers.
246. Letter of Benjamin Chew to Archibald Stewart, January 20, 1788, Chew Family Papers. On April 1, 1789, Samuel R. Hackett, of Independence, appointed Thomas Helms, merchant, of Independence, as his attorney and business agent. Thomas Helms's son, William Helms, built the first grist- and sawmill at Hackettstown. Archibald Stewart was married to William's daughter, Phoebe.
247. Helms credited Andover Forge £266 for seven tons of bar iron on March 1, 1786. On October 9, 1786, he indicated that these seven tons had sold for £174, 13 *s*, 4½ *d*. On July 3, 1787, William Helms's accounts credited Andover Forge £45 "by a Ton Nail Rod had in 1784." He also showed credit "for a Ton of Iron received at Hackettstown" on March 30, 1787. Chew Family Papers.
248. Letter from Benjamin Chew to Archibald Stewart, January 1788, Chew Family Papers.
249. Ibid.
250. Agreement between Archibald Stewart, Charles Stewart and Robert Taylor and Benjamin Chew, March 22, 1788, Chew Family Papers.
251. Chew Family Papers.
252. Letter from Benjamin Chew to Archibald Stewart, November 9, 1790, Chew Family Papers.
253. Chew Family Papers.
254. Articles of Agreement between John Armstrong and Owners of the Andover Company; Letter from Benjamin Chew to Frederick Smyth, April 1, 1788, and Copy of a Letter to John Armstrong at Andover Forge, dated Philadelphia, September 28, 1789, by Dr. Baily, Chew Family Papers. John Armstrong was born in Frelinghuysen Township on August 20, 1749, the son of Nathan and Euphemia (Wright) Armstrong. He purchased a tract of woodland near the Blue Mountains, known as the Coal Job, and erected a forge about three miles from his residence, on the west side of the Paulinskill at Paulina (formerly Lawrenceville). He made bar iron from pigs obtained at Andover Furnace until his wood gave out. Abandoning the iron business, he erected a gristmill opposite the forge, on the east side of the Kill, later owned by Cornell & Mathews. Judge John Armstrong died May 7, 1836, aged eighty-six years. See "The Death of Jacob Armstrong," *New Jersey Herald*, May 17, 1862.

255. Letter from Benjamin Chew, March 16, 1789, Chew Family Papers.
256. Chew Family Papers.
257. Letter from Benjamin Chew Jr. to John Armstrong, June 30, 1789, Chew Family Papers.
258. Chew Family Papers. Stewart apparently remained in Benjamin Chew's good graces. It is unclear whether he feigned illness to conceal chicanery or if his illness was unavoidably responsible for his lack of careful management of affairs. Archibald Stewart died at Springdale, Andover Township, in May 1795 and is buried in the Hackettstown Presbyterian Church Yard. His widow, Phoebe, died January 14, 1815.
259. Letter from Archibald Stewart, of Springdale Mills, to Robert Taylor, January 19, 1790, Chew Family Papers.
260. Indenture between Benjamin Chew and his wife, Elizabeth, of Philadelphia, and Abraham Baily, Humphrey Hill and Cadwalader Evans, of the Counties of Chester and Delaware, Pennsylvania, October 29, 1789, Chew Family Papers. Cadwalader Evans, born February 21, 1742, married Sarah Cox of Chester County, Pennsylvania, at Old Swede's Church in Philadelphia on October 11, 1778. She died in October 1794 at Andover Forge, New Jersey, and may be buried there in the ancient cemetery on the edge of the lake. Cadwalader Evans married Sarah Bond, the daughter of Colonel William Bond, of Hackettstown, New Jersey, in 1796. He moved to Brownsville, Pennsylvania, where he operated a mill.
261. Ibid.
262. Ibid.
263. Ibid.
264. Power to enter lands & tenements of John Armstrong at Andover Forge to Robert Taylor for Benjamin Chew and others, January 20, 1790, Chew Family Papers.
265. Articles of Agreement between John Armstrong and Benjamin Chew Jr. on behalf of the owners of Andover Forge, March 30, 1790, Chew Family Papers.
266. Ibid.
267. A copy of a Letter to John Armstrong carried by Mr. Cadwalader Evans, April 13, 1790, Chew Family Papers.
268. Letter from Benjamin Chew and John Lardner, May 9, 1790, Chew Family Papers.
269. Benjamin B. Edsall, "Centennial Address," in Bunnell, *Sussex County Sesqui-Centennial*, 153.
270. Futhey and Cope, *History of Chester County*, 469.
271. See Launey, *First Families of Chester County*, 12.
272. Ibid.

273. Chew Family Papers.
274. Inventory of the Goods and Chattels of Abraham Baily, Humphrey Hill and Cadwalader Evans, taken December 28, 1795, Chew Family Papers.
275. Account of Sales at Vendue of sundry items distrained at Andover Furnace, January 12, 1796, Chew Family Papers.
276. Memorandum made by Sam'l Fisher respecting the Andover Works, Chew Family Papers.
277. Letter from Amos Grandin to Benjamin Chew Jr., June 27, 1797, Chew Family Papers. On May 27, 1767, merchant John Boyton, of Philadelphia, and his wife, Elizabeth, conveyed 200 acres, known as Lot #6, to Amice [*sic*] Grandin, of Roxbury, in Morris County, for £237. Amos Grandin died in January 1817, leaving Amos Betson, son of his sister Rachel, "the obligation I hold against him." He left the children of Amos Betson the proceeds from the sale of a tract of land near the Musconetcong River, purchased from the heirs of Daniel Smith, whereon Jeremiah Cosner then lived. He left several farms on the Ayres (Eyre) Tract to Rachel Betson, Anna Batson, John Betson Jr. and Sarah Betson Young, wife of Daniel Young, with the proviso that she allow "a dam to be built by the bridge for a pond for use of the Mills where I now live." Hutchinson, NJ *Archives, Calendar of New Jersey Wills, Administrations, Etc., Vol. XIII—1814–1817*, 177. He bequeathed the tract whereon he lived to his nephews, John and Philip Grandin, and his mills to Philip, along with the right to occupy the southwest end of his dwelling house. The northeast end was reserved for his widow. The Ayres (Eyre) Tract comprised 890 acres, lying west and south of Budd Lake, originally surveyed to George Eyre.
278. Letter from Thomas Wills Jr. to Benjamin Chew Jr., June 26, 1797, Chew Family Papers.
279. *Sussex County Deed Book K*, 240.
280. Letter to Benjamin Chew Jr. from Silas Dickerson, Stanhope, March 6, 1802, Chew Family Papers. Silas Dickerson died January 7, 1807, when his great coat caught on a screw in a rapidly revolving axle and drew him into the nail-making machinery.
281. Letter from Benjamin Chew Jr. to Silas Dickerson, March 1803, Chew Family Papers.
282. Letter to Benjamin Chew Jr. from Silas Dickerson, Stanhope, October 13, 1806, Chew Family Papers.
283. Letter from Benjamin Chew Jr. to James Ludlum, Chew Family Papers.
284. Letter to Benjamin Chew Jr. from James Ludlum, of New York, regarding Andover concerns, August 31, 1807, Chew Family Papers.

THE END

285. *Acts of the Thirty-third General Assembly of the State of New-Jersey, Chapter XVII, 56–57.*
286. Chew Family Papers. Elizabeth Oswald was Benjamin Chew's second wife, the daughter of James and Mary (Turner) Oswald. Thus she was the niece of Joseph Turner.
287. *Sussex County Deed Book Z*, 173; Elizabeth Chew, widow, possessed 2½ sixteenths share, inherited of her uncle Joseph Turner, and Benjamin Chew Jr. possessed ⅛-share, acquired by his father from John Hackett.
288. *Sussex County Deed Book E2*, 407. The heirs were Margaret Morris, George Dillevyn and his wife, Sarah, and Milrah Martha Moore, all of Burlington; Gideon H. Wells and Hannah, his wife, of Trenton; William H. Wells and Elizabeth, his wife, of Dagsborough, Delaware; Benjamin W. Morris and Margaret, his wife, and Rachel H. Wells, of Wellsborough, Pennsylvania; and Richard Lamar Bissett, Edward Walsby and Henrietta, his wife, and William Davis and Mary, his wife, of the United Kingdom. George Clifford Maxwell was born in Sussex County in 1771, graduated Princeton College in 1792 and then practiced law in Flemington, New Jersey. He represented New Jersey in the House of Representatives in 1811–13 and died in 1816. Dr. Isaac Ogden, of Tewkesbury, was born at Basking Ridge in 1764. After graduating Princeton College, he eventually settled at New Germantown, working as a physician. He moved to New Brunswick in 1829, where he died.
289. *Sussex County Deed Book X*, 534.
290. Ibid., 100.
291. *Morris County Deed Book X*, 69.
292. Snell, *History of Sussex and Warren Counties, NJ*, 743. Here the name is spelled "Demund." *Sussex County Deed Book V*, 53, 117, 119. This was likely part of the original tract that Robert Willson purchased of Daniel Cox on December 17, 1753, and which Robert Willson sold to Ebenezer Willson on January 17, 1785.
293. *Sussex County Deed Book X*, 42. George Foliot (Folliott) of Chester in the United Kingdom granted 1,987.75 acres by attorney to George C. Maxwell and Andrew Bartles on June 9, 1808.
294. *Sussex County Ancient Deed Book 1*, 1.
295. Snell, *History of Sussex and Warren Counties, NJ*, 447.
296. Ibid., 468.
297. *Lake Waterloo Estates*, 16, 17. French troops did not land in Rhode Island until September 1780, months after Washington moved his army from their winter cantonment in Morristown.
298. Roy Smith erred in his belief that Indians indigenous to this area did not bury underground, as several large Native American burying grounds in the Upper

Delaware Valley, on the borders of the Paulinskill Meadows in the vicinity of Newton and Hampton and on Germany Flats in Sparta Township amply prove.

299. Sandford Roy Smith, Tape 1, Transcript, 15–16.

Bibliography

Acrelius, Israel. *A History of New Sweden; or, The Settlements on the River Delaware, Memoirs of the Historical Society of Pennsylvania*, Vol. XL. Philadelphia: The Historical Society of Pennsylvania, 1874.

Acts of the Thirty-third General Assembly of the State of New-Jersey: at a session begun at Trenton, on the twenty-fifth day of October, one thousand eight hundred and eight: being the first sitting. Chapter XVII. Trenton, NJ: James J. Wilson, 1808.

Avery, Mrs. Elroy M., ed. *The American Monthly Magazine*, Vol. XXVI, January–June 1905.

Bayley, William S. *Geological Survey of New Jersey Iron Mines and Mining in New Jersey*, Vol. VII of the Final Report Series of the State Geologist. Trenton, NJ: MacCrellish & Quigley, 1910.

Bunnell, Jacob L., ed. *Sussex County Sesqui-Centennial, September 2, 1903, Including Centennial Address of Benjamin B. Edsall.* Newton: New Jersey Herald Press, 1903.

Chew Family Papers (Collection 2050). The Historical Society of Pennsylvania.

Cook, George H. *Geology of New Jersey*. Newark, NJ: Daily Advertiser Office, 1868.

———. *Report of Professor George H. Cook Upon the Geological Survey of New Jersey and Its Progress During the Year 1863*. Trenton, NJ: David Naar, 1864.

Coxe, Tench. *A Brief Examination of Lord Sheffield's Observations on the Commerce of the United States*. Philadelphia: M. Carey, 1791.

Edsall, Benjamin B. "Centennial Address." In *Sussex County Sesqui-Centennial, September 2, 1903, Including Centennial Address of Benjamin B. Edsall*. Edited by Jacob L. Bunnell. Newton: New Jersey Herald Press, 1903.

Ford, Chauncy, ed. *Journals of the Continental Congress 1774–1789*, Vol. X. 1778, January 1–May 1. Washington, D.C.: Government Printing Office, 1908.

Forges and Furnaces in the Province of Pennsylvania. Philadelphia: Pennsylvania Society of the Colonial Dames of America, 1914.

Futhey, J. Smith, and Gilbert Cope, eds. *History of Chester County, Pennsylvania, with Genealogical and Biographical Sketches*. Vol. 2. Philadelphia: Louis H. Everts, 1881.

Geological Survey of New Jersey, Annual Report of the State Geologist for the Year 1899, Report on Forests. Trenton, NJ: MacCrellish & Quigley, 1900.

Gillespie, Charles, ed. *A Diderot Pictorial Encyclopedia of Trades and Industry, 485 Plates Selected from "L'Encyclopédie" of Denis Diderot*. New York: Dover Publications, Inc., 1959.

Goodrich, Charles A., ed. *A New Family Encyclopedia or Compendium of Universal Knowledge*. Philadelphia: Charles A. Goodrich, 1833.

Hall, John. *History of the Presbyterian Church in Trenton, N.J.* New York: Anson D.F. Randolph, 1859.

Hermelin, Samuel Gustaf. *Report About the Mines in the United States of America 1783*. Philadelphia: John Morton Memorial Museum, 1931.

Honeyman, A. Van Doren, and William Nelson, eds. *Archives of the State of New Jersey. First Series, Volume XXIX, Documents Relating to the Revolutionary History of the State of New Jersey, Extracts from American Newspapers Relating to New Jersey Vol. X, 1773–1774*. Paterson, NJ: Call Printing and Publishing Co., 1917.

Honeyman, A. Van Doren, ed. *Archives of the State of New Jersey, First Series Vol. XXXI, Extracts from American Newspapers Relating to New Jersey, Vol. XI, 1775*. Somerville, NJ: Unionist-Gazette Association, Printers, 1923.

———. *Archives of the State of New Jersey. First Series, Volume XXXIII, Documents Relating to the Revolutionary History of the State of New Jersey, Calendar of New Jersey Wills, Administrations, Etc., Volume IV, 1761–1770*. Somerville, NJ: The Unionist-Gazette Association, Printers, 1928.

Hunt, Gaillard, ed. *Journals of the Continental Congress 1774–1779*, Volume XVI, 1780 January 1–May 5. Washington, D.C.: Government Printing Office, 1910.

Hutchinson, Elmer T., ed. *Archives of the State of New Jersey First Series, Vol. XLII, Documents Relating to the Colonial, Revolutionary and Post-Revolutionary History of the State of New Jersey, Calendar of New Jersey Wills, Administrations, Etc., Vol. XIII—1814–1817*. Trenton, NJ: MacCrellish & Quigley Co., 1949.

Kemper, Jackson. *American Charcoal Making*. Washington, D.C.: Eastern National Park and Monument Association, United States Department of the Interior, National Park Service, 1941.

Kimball, David A., and Miriam Quinn. "William Allen—Benjamin Chew Correspondence, 1763–1764." *The Pennsylvania Magazine of History and Biography* 90, no. 2 (April 1966).

Kistler, Ruth Moser. "William Allen, Founder of Allentown." *Proceedings of the Lehigh County Historical Society* 24 (1962).

Kitchell, William. *Third Annual Report on the Geological Survey of the State of New Jersey for the Year 1856.* Trenton, NJ: True American Office, 1857.

Lake Waterloo Estates. Newark, NJ: Colyer Printing Co., 1929.

Launey, John Pitts. *First Families of Chester County, Pennsylvania.* Vol. 2. Westminster, MD: Willow Bend Books & Family Line Publications, 1999.

Lesley, J.P. *The Iron Manufacturers' Guide to the Furnaces, Forges and Rolling Mills of the United States.* New York: John Wiley, 1859.

Minutes of the Supreme Executive Council of Pennsylvania from Its Organization to the Termination of the Revolution. Vol. XV. Harrisburg, PA: Theodore Fenn & Co., 1853.

Moldenke, Richard. *The Principles of Iron Founding.* New York: McGraw-Hill Book Company, Inc., 1917.

Mulholland, James A. *A History of Metals in Colonial America.* Tuscaloosa: University of Alabama Press, 1981.

Nelson, William, ed. *Archives of the State of New Jersey, First Series, Documents Relating to the Colonial History of New Jersey. Vol. XIX, Extracts From American Newspapers Relating to New Jersey Vol. III 1751–1755.* Paterson, NJ: Press Printing & Publishing Co., 1897.

———. *Archives of the State of New Jersey, First Series Vol. XX, Extracts from American Newspapers, Relating to New Jersey, Vol. IV. 1756–1761.* Paterson, NJ: Call Printing and Publishing Co., 1898.

———. *Archives of the State of New Jersey. First Series Vol. XXIV, Documents Relating to the Revolutionary History of the State of New Jersey, Extracts from American Newspaper Relating to New Jersey, Vol. V 1762–1765.* Paterson, NJ: Press Printing and Publishing Co., 1901.

———. *Archives of the State of New Jersey. First Series Vol. XXV, Documents Relating to the Revolutionary History of the State of New Jersey, Extracts from American Newspapers Relating to New Jersey, Vol. VI, 1766–1767.* Paterson, NJ: Call Printing and Publishing Co., 1903.

———. *Archives of the State of New Jersey, First Series Vol. XXVII, Extracts from American Newspapers, Relating to New Jersey, Vol. VIII. 1770–1771.* Paterson, NJ: Press Printing and Publishing Co., 1905.

———. *Archives of the State of New Jersey, First Series Vol. XXVIII, Extracts from American Newspapers, Relating to New Jersey, Vol. IX, 1772–1773.* Paterson, NJ: Call Printing and Publishing Co., 1916.

Overman, Frederick. *The Manufacture of Iron in All Its Various Branches.* Rev. 3rd ed. Philadelphia: T.K. Collins and P.G. Collins, 1854.

Papers of the Continental Congress. National Archives.

Prince, Carl E., and Dennis P. Ryan, eds. *The Papers of William Livingston, Vol. 2: July 1777–December 1778.* Trenton: The New Jersey Historical Commission, 1980.

Prince, Carl E., Dennis P. Ryan, Brenda Parnes and Mary Lou Lustig, eds. *The Papers of William Livingston, Vol. 3: January 1779–June 1780.* New Brunswick, NJ: Rutgers University Press, 1986.

Proceedings of the New Jersey Historical Society, Vol. VII, 1853–1854, No. 2. Newark, NJ: Daily Advertiser Office, 1855.

Schultz, George W. *Antique Iron Works and Machines of the Water Power Age*. Bowers, PA: Geo. W. Schultz, 1927.

Scott, Austin, ed. *Archives of the State of New Jersey. Series Two, Volume V, Documents Relating to the Colonial History of the State of New Jersey, Extracts from American Newspapers Relating To New Jersey, October, 1780–July, 1782*. Trenton, NJ: State Gazette Publishing Co. Printers, 1917.

Second Annual Report on the Geological Survey of the State of New Jersey, for the Year 1855. Trenton, NJ: True American Office, 1856.

Selections from the Correspondence of the Executive of New Jersey, From 1776 to 1786. Newark, NJ: Newark Daily Advertiser, 1848.

Sims, P.K., and B.F. Leonard. *Geology of the Andover Mining District, Sussex County, New Jersey*. Geologic series, issue 62. Trenton, NJ: Division of Planning and Development, 1952.

Smith, Robert F. "A Veritable…Arsenal of Manufacturing: Government Management of Weapons Production in the American Revolution, A Dissertation Presented to the Graduate and Research Committee of Lehigh University in Candidacy for the Degree of Doctor of Philosophy in the Department of History." Easton, PA: Lehigh University, 2008.

Smith, Sandford Roy. Tape 1, Transcript (1980).

Snell, James P., comp. *History of Sussex and Warren Counties, New Jersey*. Philadelphia: Everts & Peck, 1881.

Stryker, William S., ed. *Archives of the State of New Jersey, Second Series, Documents Relating to the Revolutionary History of the State of New Jersey, Second Series, Vol. I, Extracts from American Newspapers Relating to New Jersey, 1776–1777*. Trenton, NJ: John L. Murphy Publishing Co., 1901.

Tenney, William J., ed. *The Mining Magazine: Devoted to Mines, Mining Operations, Metallurgy, &c.* Vol. 1. New York: John F. Trow, 1853.

"Three maps of northern New Jersey, with references to the boundary between New York and New Jersey" [1769?]. *William Faden's Catalogue of a Curious and Valuable Collection of Original Maps and Plans* [Boston, 1862], Library of Congress, Maps of North America, 1750–1789, 1245. (Digital ID: g3811far124500 http://hdl.loc.gov/;oc.gmd/g3811f.ar124500).

Tiemann, Hugh P. *Iron and Steel (A Pocket Encyclopedia)*. New York: McGraw-Hill Book Company, Inc., 1919.

Walker, Lewis Burd. *The Burd Papers, Extracts from Chief Justice William Allen's Letter Book*. Pottsville, PA: Lewis Walker Burd, 1897.

Webb, Edward A. *Sussex County, N.J. Historical Directory*. Andover, NJ: Edward A. Webb, 1872.

Whitehead, William A., ed. *Archives of the State of New Jersey, Documents Relating to the Colonial History of the State of NJ, Vol. VII, Part of the Administration of Governor Jonathan Belcher, 1746–1751*. Newark, NJ: Daily Advertiser Printing House, 1883.

Wilson, Thomas B., ed. *Notices from New Jersey Newspapers 1781–1790*. Lambertville, NJ: Hunterdon House, 1988.

Index

C

D

E

F

G

T

Z

About the Author

For thirty-six years and counting, Kevin Wright has worked to provide meaningful and memorable experiences to a diverse audience at some of New Jersey's most storied places. Upon graduating from Rutgers College, he began his career in the water-powered gristmill at Waterloo, New Jersey, and was soon promoted to tour director of the restored village. In 1981, he and his young family moved into the Steuben House on the Revolutionary War battleground at Historic New Bridge Landing in River Edge, where he was state curator and historical interpreter for twenty years. He worked eight years as northern regional resource interpretive specialist. He was integral to the visioning process for Historic New Bridge Landing, the Newton Town Plot Historic District, Lusscroft Farms and the Spirit of the Jerseys State History Fair. Long devoted to research and writing, he has become a noted author and speaker on a variety of historical topics. The tenth generation in his father's lineage to be born in Sussex County, New Jersey, and the fourth generation on his mother's side to live in Bergen County, he has served in many capacities as a volunteer and preservation advocate, including as president of both the Sussex County and Bergen County Historical Societies.